Can Microbial Communities Regenerate?

CONVENING SCIENCE

DISCOVERY AT THE MARINE BIOLOGICAL LABORATORY

A Series Edited by Jane Maienschein

For well over a century, the Marine Biological Laboratory (MBL) has been a nexus of scientific discovery, a site where scientists and students from around the world have convened to innovate, guide, and shape our understanding of biology and its evolutionary and ecological dynamics. As work at the MBL continuously radiates over vast temporal and spatial scales, the very practice of science has also been shaped by the MBL community, which continues to have a transformative impact the world over. This series highlights the ongoing role MBL plays in the creation and dissemination of science, in its broader historical context as well as for current practice and future potential. Books in the series will be broadly conceived and defined, but each will be anchored to MBL, originating in workshops and conferences, inspired by MBL collections and archives, or influenced by conversations and creativity that MBL fosters in every scientist or student who convenes at the Woods Hole campus.

Can Microbial Communities Regenerate?

UNITING ECOLOGY AND EVOLUTIONARY BIOLOGY

S. Andrew Inkpen
and W. Ford Doolittle

The University of Chicago Press

Chicago and London

The University of Chicago Press, Chicago 60637
The University of Chicago Press, Ltd., London

Published 2022
Printed in the United States of America

31 30 29 28 27 26 25 24 23 22   1 2 3 4 5

ISBN-13: 978-0-226-82063-7 (cloth)
ISBN-13: 978-0-226-82034-7 (paper)
ISBN-13: 978-0-226-82035-4 (e-book)
DOI: https://doi.org/10.7208/chicago/9780226820354.001.0001

Library of Congress Cataloging-in-Publication Data

Names: Inkpen, S. Andrew, author. | Doolittle, W. Ford, 1942– author.
Title: Can microbial communities regenerate? : uniting ecology and evolutionary biology / S. Andrew Inkpen and W. Ford Doolittle.
Other titles: Convening science.
Description: Chicago : University of Chicago Press, 2022. | Series: Convening science: discovery at the Marine Biological Laboratory | Includes bibliographical references and index.
Identifiers: LCCN 2021059134 | ISBN 9780226820637 (cloth) | ISBN 9780226820347 (paperback) | ISBN 9780226820354 (ebook)
Subjects: LCSH: Regeneration (Biology) | Microbial populations. | Evolution (Biology)
Classification: LCC QH499 .I55 2022 | DDC 571.8/89—dc23/eng/20211217
LC record available at https://lccn.loc.gov/2021059134

♾ This paper meets the requirements of ANSI/NISO Z39.48-1992 (Permanence of Paper).

To Mary's Place Cafe II

Contents

1 Regeneration

If there were no regeneration there could be no life.
If everything regenerated there would be no death.
RICHARD GOSS, *Principles of Regeneration*, 1969[1]

This book is about regeneration—what it is, why it matters, why it occurs, what we can do to foster it, and how it compares across the nested hierarchy of biological entities we call life. Regeneration is a ubiquitous feature of living systems and, in general, too large a topic for any single short book. So we frame our investigation here around one of the smaller scales in the biological hierarchy, the microbial scale, focusing specifically on the regeneration of communities of microorganisms: collections of different kinds of microbes interacting with one another at a particular location, such as within the human large intestine. The regeneration of microbial communities more generally is crucial, however, because microbes support all life, from multicellular organisms like humans to entire ecosystems. During the first two-thirds of the history of this planet, microbes *were* all life, and they are most of it now!

FIGURE 1.1 | A sixteenth-century engraving of Hercules in battle with the Lernean Hydra, by the famous Dutch engraver Cornelis Cort (1533–1578). (Photo: Wikimedia Commons.)

Humans have long been fascinated by the idea of regeneration, perhaps because, as biological organisms, we seem to be comparatively poor at it. It was the quest to understand regeneration in certain invertebrates and salamanders, both able to regenerate lost limbs or other major body parts, that inspired the dissections and experiments of eighteenth-century naturalists, today recognized as the forebears of modern experimental biology.[2] If only we humans could understand and harness this mysterious ability! And long before it became a topic for scientific study, regeneration appeared in many myths. Hercules fought the Hydra, a serpent that frustratingly regenerated two heads in the place of one chopped off.[3] To best his foe, Hercules cauterized the stumps of its heads with fire, thereby stalling their regeneration (fig. 1.1).

But regeneration goes far beyond salamander limbs and mythical heads, beyond the regrowth of an organism's organs or parts. It occurs as living systems at all scales, from cells to ecosystems, *maintain, recover,* or *reproduce* themselves in the face of continual damage and disturbance. Without regeneration, many biological systems would be too fragile to cope with stress and would collapse. In an age when anxiety about sustainability and environmental collapse has become more than science fiction, understanding regeneration has taken on new significance.

Examples are the best proof of the startling regenerative diversity that exists in the organic world. Expert organism regenerators, like *Hydra*—the tiny freshwater organism, not the mythical serpent—have the ability to regenerate their entire bodies if cut into many pieces. Because this regeneration occurs without normal cell division, the resulting regenerated *Hydra* is a cute, tiny version of its "parent." Closer to home, many human tissues also have the ability to regenerate. Our skin continually regenerates through cell division as skin cells slough off from daily wear and tear. Our intestinal mucosal cells and red blood cells, too, continuously regenerate. We're not hydras, perhaps, but we do not lack regenerative capacity altogether. At a much larger scale, a Canadian boreal forest can, under the right conditions, naturally regenerate after a forest fire through the establishment of species adapted to such conditions, such as the lodgepole pine. This pine produces cones protected by a waxy coating that requires the heat of a fire to release their seeds. And in a sense, even species themselves regenerate through the reproduction of organisms.[4]

One wonders, after encountering such a miscellany, whether anything might unify how regeneration works across biological scales—and Marine Biological Laboratory scholars Kate MacCord and Jane Maienschein address this question in *What Is Regeneration?*, the first book in this series, arguing that a systems-based approach provides a promising avenue.[5] Our aim in this book is to build on their approach but focus on the regeneration of one kind of biological system—the *microbial community*—and look at it through the lenses of ecology and evolutionary biology. Drawing on historical and current biological research in microbiology and contemporary philosophy of biology, our goal is to offer a way of understanding how microbial community regeneration works and why it occurs. This turns out to be a timely topic at several levels. Many are concerned about the best diet to adopt if the microbes in their guts are to work harmoniously after being disrupted by disease or antibiotic use. And many, too, are concerned about whether the Earth will recover after the loss of much of its biodiversity, a recovery that depends on the regeneration of microbial communities.

The idea of regeneration in general might appear a dizzyingly complex phenomenon, occurring at many different biological scales and over different timescales. Matters are made worse by the fact that researchers have not always used "regeneration" to mean the same thing. Regeneration has been variously treated as synonymous with a number of other re- words, as Maienschein and MacCord note: re-juvenation, re-vitalization, re-newal, re-mediation, re-pair, re-storation, re-plication, re-covery, re-placement, and re-silience. Some of these terms emphasize the original system or starting state before disturbance; some emphasize the end state, and that it is beneficial for the living system or adaptive; some seem to emphasize a "normal state" against which to assess injury or damage. As we begin to think deeply about regeneration, our lack of clarity is apparent and threatens the usefulness of the concept. This book is aimed at providing helpful precision where there is currently unhelpful obscurity.

To render thinking about regeneration manageable, it is helpful to split regeneration into two types. The first relates to individual organisms and their parts, like the regeneration of the salamander limb or the human liver. We will call this *organismal regeneration*. Long-standing problems of organismal regeneration include: Why can some organisms regenerate and not others? Why can some parts of some organisms regenerate and not other parts? Why have many organisms lost the ability to regenerate over evolutionary time?

The second type relates to living systems called collectives, systems consisting of more than one individual organism. These include populations (single-species collectives), symbioses (two-species collectives), and our focus in this book, *communities* (multispecies collectives). Communities are the primary subject of the science of ecology. When ecologists speak of the regeneration of a Canadian boreal forest, or microbiologists speak about recovery from *Clostridium difficile* infection in humans, they are speaking of collective regeneration: the regeneration of the multispecies community of organisms distinctive of such forests or the human intestinal tract. Often the word "recovery" (not "regeneration") is used in these contexts. And, as explained later in this book, it is often the pattern of functional relationships between species, not (or not necessarily) the very same collection of species that is recovered or restored. We consider this to be regeneration, too.[6]

While the distinction between organismal regeneration and collective regeneration is intuitive and helpful, the biological world, as always, pushes back against our endeavor to achieve such conceptual tidiness. Is a honeybee colony a collective or an organism in its own right? What about colonial organisms, such as the jellyfish-like Portuguese man-of-war: the different zooids that constitute the parts of a man-of-war—its peculiar version of separate "organs"—act independently enough that they are judged to be separate but cooperating colonies. Rather than take a firm stand on this vexed topic of what counts as an individual organism, we will allow that some biological systems fall into both categories and insist that our focus will be on *communities* (whether or not they also qualify as organisms).

The concept of community regeneration appears to have its origin much later than did that of organismal regeneration, arising first in nineteenth-century discussions of the regrowth of Royal Forests in France and England.[7] By the early twentieth century, following the emergence of ecology as a scientific discipline, regeneration became a household word to refer to the reemergence of a similar community of organisms following some disturbance, such as a fire or logging. Frederic Clements (1874–1945), a botanist at the Carnegie Institution of Washington and a foundational figure in ecology whom we

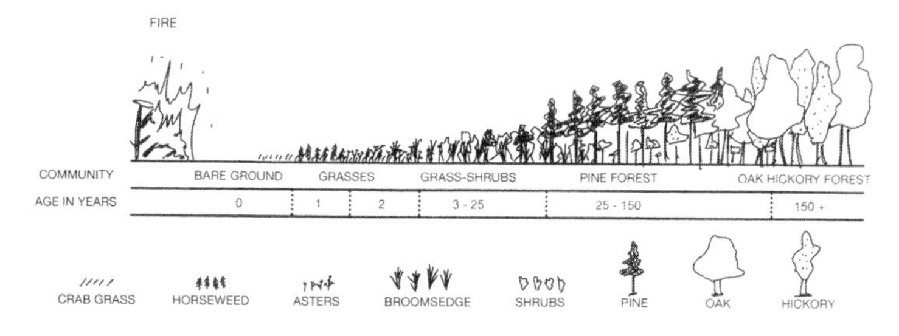

FIGURE 1.2 | Ecological succession, from barren ground to oak-hickory forest, such as might occur in the Georgia piedmont.

discuss further in the chapter 2, gave the phenomenon its first theoretical foundation.

Clements thought that a community *was* a type of organism and, like organisms, communities also possessed the capacity for regeneration. Indeed, he referred to communities as *super-organisms*: organisms made up of other organisms.[8] But, influenced by the success of embryology, his thinking went further than this: he also reasoned that plant communities develop in the way that organisms do—what he termed "ecological succession" was a process analogous to organismal development. Just as an organism develops from an unfertilized egg, through a number of differently organized stages, ending at an adult individual, so too with a plant community. Starting with barren ground, Clements reasoned, a sequence of predictable stages of community composition unfold until the mature, or "climax," community is achieved (fig. 1.2). For regeneration, he coined the phrase "secondary succession": the restorative process that occurs when a community is damaged.

Clements's vision of community regeneration was bold and went considerably beyond what was justified by his data—and beyond more modern conceptions of organisms. He believed regeneration was *progressive* in that the community became more complex and more resistant to damage over time. It was also goal-directed or *teleological*: the process invariably led to the same climax community—it seemed to have that "goal" or "purpose" in mind, so to speak, from the start. And it was highly *predictable*: as long as the climate and soil type remained the same, the same communities would develop in the same geographic areas, over and over.

Modern ecologists now consider Clements's theory of succession to be problematic—ecologist Frank Egler once offered a $10,000 reward to any ecologist who could provide evidence in its favor.[9] Still, the idea has considerable cultural influence and a modified theory of succession maintains a privileged place in ecology and in theories of community regeneration. Moreover, the idea of the superorganism is making a revival in contemporary discussions of what are called holobionts: these are organisms that consist of an animal or plant or protist "host" and the microorganisms commonly associated with them. For example, a human and the many microbes we harbor throughout our bodies may together constitute a holobiont. Thus, ultimately, even though he knew almost nothing of microbes, and even though he took a few wrong turns, Clements provided a theoretical foundation similar to that on which we will build throughout this book.

We said above that regeneration in ecology has become one name for the reemergence or restoration of a similar ecological community following a disturbance. This is a good start, but part of the intended significance and novelty of this book is in rendering this common idea more precise. With this goal in mind, we flag two significant features of this definition of regeneration that are worthy of further discussion. First, it is backward-looking, in that regeneration involves something from the past reemerging. This, however, may appear to leave out something very important about the idea of regeneration, namely, that it is beneficial or adaptive. When we say that a community has regenerated, we often mean more than just that is has reemerged. We often imply the idea of repair. We imply that regeneration has been good for the system. But what does it mean for something to be good for a community over-and-above what is good for the individual organisms that compose it? Should we drop this idea of what is "good for the community" altogether or can biological sense be made of it? Is it the same as how the regeneration of a limb is good for an organism, like a salamander? Accounting for apparent goal-directedness or purposiveness, ascribing benefit, and understanding the causes of adaptedness are old problems in biology, in no way unique to regeneration, and we return to these in detail in later chapters.

The second feature of this definition of regeneration that is worthy of further discussion is the idea of an ecological community. Defining what counts as a community has been a controversial subject among ecologists and philosophers, with heated debates tracing all the way back to the origins of ecology in the early twentieth century. A community is most often said to be a collection of organisms from different species living in the same area, but this often needs to be qualified. One qualification, important for us in this book, has to do with how bacteria are classified. We are discussing the microbial world, and the dominant understanding of species coming from traditional studies of animals (birds, often) is that a species is a group of actually or potentially interbreeding organisms. This does not apply to organisms which often reproduce asexually, like bacteria. For bacteria, only one parent is required to create offspring, and thus "interbreeding" does not necessarily occur. Many bacteria do exchange genes by recombination but very much less frequently than the once per generation which we and many animals are obliged to do by our sexual reproductive mechanisms. Bacteria seldom "mate," that is, and never by this mechanism. Thus, it is an open question whether bacteria even "have" species. This fits awkwardly with the common definition of community as a multi*species* collective. Many microbiologists, in order to avoid these issues, use operative taxonomic unit or "phylotype," defined as a cluster of organisms grouped by genetic relatedness as measured by similarity in DNA sequence, as a substitute for the word "species." Throughout this book, though, we will often speak of bacterial species as a shorthand.[10]

A different qualification relates to how we should understand what makes communities similar to one another or of the same type—that is, how we classify communities. This matters for regeneration because it is said to occur with the reemergence of a similar community following a disturbance. But what is meant by "similar"? Traditionally, there are two different interpretations. One is what we call "taxonomic"; the other is "functional." Consider first an analogy. Down the street from one of our houses there once was a French cafe that served delicious croissants as well as other pastries, sandwiches, and coffee. Sadly, this cafe closed but, fortunately, another cafe took its place. The new cafe, which is German rather than French, offers a very different set of pastries and sandwiches than the previous cafe and makes a different cup of coffee. If we define these cafes taxonomically, according to the composition of their baked goods, then a very different cafe has emerged where the old one stood—the new cafe has *Franzbrötchen* where the croissants once lived. But if, on the other hand, we define the cafes functionally, according to the types of food offered, they both offer pastries and sandwiches and coffee, and we might say that, functionally speaking, a very similar cafe exists now where the old one stood.

Turning to ecology, taxonomically, a straightforward way to define a community is by making a catalog of its member species. When we said that a Canadian boreal forest can regenerate following a forest fire or a gut recovers from a regimen of antibiotics, we might mean, following this taxonomic definition of community, that the same species came to occupy the area after regeneration as occupied it previously. Alternatively, we can define a community functionally, as the set of ecological roles played by its members. An example, discussed further below, will help here. Humans harbor a community of microorganisms in their guts that can beneficially contribute to physiological processes like digestion. And different healthy humans harbor taxonomically different communities, that is, they contain different bacterial species. In fact, if we take antibiotics which affect the composition of our gut communities by killing bacteria, then later but still healthy versions of ourselves may harbor a different community, taxonomically speaking. But as long as this community has bacteria playing the same kinds of roles in the community, thus helping us out with digestion, we may say that we have successfully regenerated our gut communities from the functional point of view. Sometimes, as in the case of chronic *Clostridium difficile* infection, therapeutic interventions as discussed in detail in chapter 5 might be required to bring about regeneration. Throughout this book, we discuss examples of both *taxonomic* and *functional* community regeneration, making it clear which we mean.[11]

A survey of today's science of microbiology and its applications gives the impression that the ultimate goal is to bring about functional, rather than taxonomic, regeneration. This is even encouraged by the technologies biologists employ to learn about microbial communities. In these days of cheap and easy gene sequencing, functional community regeneration is inferred indirectly from a community's content of relevant genes rather than any direct assessment of those genes' activities. That is, when scientists discover that a gene is present, they infer also that the metabolic activity usually provided by that gene (and known from previous studies) is also present. The new science of "metagenomics" even encourages sequencing all of a community's genes (or the messenger RNAs they produce) without assigning them to any particular species that might be present. Indeed, two sites, or the same site after disturbance, might be said to be functionally similar as long as they possess similar metabolic activities (again, as inferred indirectly from the genes present) regardless of the species in which those genes previously evolved and are currently housed. It's often function, not taxonomy, that matters.

It is worth noting that there is a parallel to this situation in organismal regeneration occurring at the developmental rather than the genetic level. For example, consider the lamprey, a species of jawless fish. When the spinal neuron of a lamprey is severed, it can regenerate and restore the fish's swimming ability. But the structure of the neuron and the developmental path it follows are not the same after regeneration as before the injury.[12] So here, too, we have functional regeneration without perfect structural fidelity. Developmental biologist Richard Goss rightly noted back in the 1960s that "regeneration is to be regarded primarily as a device by which functional competence is recovered. Morphological restitution is only a means to this end."[13] The scientists working in metagenomics today often think similarly: regeneration occurs when community functions are recovered regardless of which species help perform those functions—the species are merely a means to an end.

We've now said a bit about what regeneration is and how we define a community. Let us turn to what we mean by *microbial.* During much of the twentieth century, discussions of regeneration in ecology focused on "macrobial" organisms. These are the everyday organisms we can see with the naked eye, things like oak trees, mice, elephants, ants, or salmon. In this book, our focus is on microorganisms, the word biologists use to refer to life-forms, mostly unicellular, of a few different types: prokaryotes (bacteria and archaea), protists, and unicellular fungi and algae. Microbial collectives are less easily visible, typically requiring microscopes or other technological help, but no less consequential for our daily lives. Dental plaque, that stuff that grows on your teeth and that prompts anxiety-inducing visits to the dentist, consists of a community of greater than 700 bacterial species in a slimy matrix called a biofilm.[14] Archaea seem seldom to cause disease and protists are generally rarer than bacteria, so bacteria have been the main focus of recent science and also, for this reason, will be the main focus of this book.

The field of microbial ecology dates to the early twentieth century, but as a result of recent advancements in DNA-sequencing technologies that make microbes easier to study, this field has sprung to the forefront of modern biology.[15] Now understanding microbial collectives and their restoration (and as we will argue later, regeneration) sits at the basis of both planetary sustainability and human medicine. Three examples will help to make this point.

From a sustainability standpoint, "the microbial world constitutes the life support system of the biosphere," as a 2019 *Nature* Consensus Statement put it, and understanding how microbial communities will respond to human-induced climate change is a major focus of current research.[16] For example, communities of marine phytoplankton, microorganisms like bacteria and algae that can perform photosynthesis, serve as the basis of the marine food web and provide an essential, natural long-term mechanism for sequestering greenhouse gases from the atmosphere. They perform about half of the world's carbon fixation (as much as all terrestrial plants combined!) while amounting to only 1 percent of total plant biomass. Moreover, they produce about half the world's oxygen.

Understanding how microbial communities will respond to human activities such as greenhouse gas emissions, pollution, and ocean acidification is an area of critical research importance. Like a complex chemical reaction, the end products of microbial metabolism (what is produced when microorganisms "eat") as well as microbes themselves constitute the starting reagents that drive the continuing maintenance of living systems at higher biological levels and, thus, the resilience of the entire ocean biosphere.

A second example demonstrates the deep connection between *micro*bial community regeneration and *macro*bial. Like other animals, the honeybee harbors a diverse microbial community within its gut, containing a mix of beneficial (helpful), pathogenic (harmful), and commensal (neither beneficial nor harmful) microbes.[17] Each honeybee gut is like an ecological island. Newly emerged worker bees are bacteria-free but are colonized shortly after birth. These bacterial colonists immigrate from the bee's environment, largely from the guts of other bees in the hive as the bees socialize with one another. The composition of a bee's gut is important because the resident microbes determine its abilities to digest food, regulate its immune system, and defend against pathogens. Bees experimentally deprived of their normal microbiota show reduced weight gain, increased susceptibility to pathogens, and increased mortality. There is also evidence that microbes help bees recognize their colony mates and distinguish them from unrelated bees.[18] Individual bee health percolates upward. The health of the bees in a hive affects the health of the colony and the ability of the hive to regenerate. And because bees are major pollinators, they are essential for plant community regeneration, which in turn affects the regeneration of animal communities. Ensuring that the microbial community in the gut of the bee regenerates successfully is about more than just microbes and bees.

A final example relates specifically to human medicine. Microorganisms have long been a focus of medical research and treatment—just think of the antibiotics used to treat bacterial infections or of the well-known scientists Louis Pasteur (1822–1895) and Robert Koch (1843–1910) whose discoveries initiated medical microbiology. For much of the history of medicine, microbes were treated as hostile invaders that medicine should wage a war against—but this attitude is changing.[19] Research has shown that many microbes are essential to the normal physiological processes of many macrobial organisms (animals and plants), their presence being imperative for health and well-being.[20] We humans host helpful communities of microorganisms—components of what in the medical context are often called "microbiomes"—in and on our bodies. A popular example introduced above is the gut microbiome, the community of microorganisms harbored in the human intestinal tract.[21] In healthy human adults, the activities of this microbial community are a part of fundamental physiological processes of the human host, such as the metabolic functions of digestion and vitamin production. Antibiotic treatment, often aimed at treating infection elsewhere in the body, can disrupt the functions of this microbial community, which can be detrimental to the health of the human host.[22] Following treatment, and in a healthy adult human, the microbial community can regenerate, and once again provide vital functions. Like sustainability, much medical research is now focused on understanding the regeneration of microbial communities. This is part of the new science known as "microbiomics," which was spawned by technological advances in our abilities to study genetic material, known as "metagenomics" (as discussed above).

We would like to help such efforts succeed. But how can we help with this book? We see three ways. First, as we've said above, we aim to stand back, so to speak, from each of these specific examples, to see the forest from the trees, and offer a way of *explaining* microbial community regeneration. This requires both biology and philosophy as we attempt to distinguish throughout those questions that are straightforwardly answered by doing experiments and collecting data from those that are informed by assumptions that are more philosophical in nature (for instance, what even *is* an organism?). Let us say a little more about this goal now.

In general terms, explanations provide successful answers to how-questions and/or why-questions. If you ask one of us how he bought the last *Franzbrötchen* at the local cafe, he might say he used his credit card, but the answer to why is that he was hungry for sugar. In explaining microbial community regeneration, we offer answers to both *how* regeneration occurs and also *why* it occurs. Following the biologist-turned-historian Ernst Mayr, biologists and philosophers today typically distinguish between these types of explanation using the labels "proximate" (how-) explanations and "ultimate" (why-) explanations. Mayr himself wrote that the evolutionary biologist, impressed by the enormous diversity of the organic world, "wants to know the reasons for this diversity [the why] as well as the pathway by which it has been achieved [the how]."[23] After describing these types of explanation in the immediately preceding paragraphs, we will return to this distinction throughout our future discussions in this book.[24] This is a helpful distinction, we would say imperative, because it allows us to clarify how different ways of explaining regeneration relate to each other and it demonstrates how they can be complementary rather than conflicting. Moreover, along with another distinction we draw in chapter 3 (that between "accidental/automatic" and "selected for" regeneration), it allows us to develop and further clarify how organismal and community regeneration, specifically, relate to each other. As we will see in chapter 2, this has long been a vexed issue: once Clements's superorganism theory came under attack and ecologists rejected the idea that communities were organisms (or even organism-like), a theoretical wedge was driven between explaining the capacities of *organisms* and the capacities of *communities,* including the capacity for regeneration.

Turning to the distinction itself, a proximate explanation is one that appeals to the sequence of causes that brings something about. Again, most naturally put, these causes explain *how* something came about. Before turning back to the biology, consider a further question we could ask at the local cafe: *how* did my paper coffee cup come to have a corrugated cardboard sleeve? A successful proximate explanation, drawing attention to the causes that bring about this state of affairs, is that the barista puts sleeves on the cups before filling them with coffee and handing them across the counter. We will sometimes refer to these causes as *proximate* causes.

An ultimate explanation, in contrast, according to our use in this book, appeals to the function something performs or its purpose (its so-called *ultimate* or final cause).[25] Most naturally, appealing to its function or purpose explains why something is the way it is. We might also wonder: why do paper coffee cups have corrugated cardboard sleeves? A successful ultimate explanation, drawing attention to the function or purpose of coffee sleeves, is that sleeves prevent piping hot coffee from burning our hands. That is why they are there.

These are, of course, what philosophers call toy examples, analogies not necessarily derived from the "real world" they are meant to exemplify. We're not really interested in explanations offered at the local cafe but in the how and why of microbial community regeneration. We will consider the proximate explanation of community regeneration in detail in the chapter 2 and this will involve a discussion of ecology. We address the question: how does microbial community regeneration occur? We show that, proximately speaking, regeneration occurs as a result of ecological interactions among members of a community that are described through a theory called "community dynamics" (basically, the rules that summarize change in the composition of a community over time). To understand how (and whether) regeneration will occur, one needs to know about the initial disturbance (for example, how badly and in what way did the antibiotics you took affect the community of microbes living in your gut), the composition of the wider species pool (that is, what microbes are close enough by to rejoin the community), and the ways that microorganisms interact with one another (that is, who gets along with who, and in what ways do they "get along").

But biology is an interesting science because this is sometimes not the only type of explanation one can provide. We also need to discuss the ultimate explanation of microbial community regeneration, why it occurs, and in our discussion that follows this will involve appealing to its function or purpose. We discuss that ultimate explanation in the latter half of the book.

Although it may sound problematically anthropomorphic or even metaphysically spooky to invoke functions or purposes to explain natural phenomenon without intentions or desires, there is a way that the theory of evolution by natural selection can render such speech, well, natural. Simply put, this theory explains the form of current living creatures by appealing to the history of the differential survival and reproduction of their ancestors (in the biological vernacular: the historical outcome of heritable variation in fitness differences). Philosophy of biology can help to sort out how this theory works to endow living forms with functions and purposes. Consider a different set of how and why questions: how and why do turtles have shells? An answer to the how-question, a proximate explanation, explains this phenomenon by appealing to the turtle's development and life history. They have shells because certain genes activated early in development eventually, through a set of complicated causal pathways, create a shell. An answer to the why-question, in contrast, would explain the phenomenon by appealing to the evolutionary history of turtles and show that turtles have shells because shells help protect the turtles from being eaten by predators. That is the function or purpose of the turtle's shell, and it evolved gradually, generation after generation, with each generation having better protection than the last. See, nothing spooky. The evolutionary function or purpose of a biological phenomenon is simply that of its effects that was naturally selected, so to speak.

That's all well and good for turtle shells, but we want to understand why microbial community regeneration occurs—what is its evolutionary function or purpose? Does it have one at all? Our question, then, considered in chapter 3, is whether evolutionary thinking can be applied to microbial communities as it can to turtles. If it can, then microbial community regeneration may have an evolutionary function or purpose, and an ultimate explanation would then seem appropriate. If evolutionary thinking cannot be applied to microbial communities, then it might have only a proximate explanation; a how, but no why, to use our terminology. Applying evolutionary theory to communities is an old problem in biology, and it turns out not to be so easy from a more conservative Darwinian evolutionary perspective. So that may be bad news. But, in chapter 4, we propose some workarounds—the good news!

Beyond evolution and natural selection, there is also a different way that ultimate explanation enters our discussion of regeneration. When living systems are designed and engineered by humans for a particular function or purpose, an ultimate explanation is also warranted, although not an evolutionary one. We will call these engineering ultimate explanations and distinguish them from the evolutionary ones discussed in the last paragraph. Instead of appealing to a history of natural selection, we appeal to human design and intention; in either case, we appeal to the purpose or function in order to explain why something is the way it is.

For example, we might ask: why do monocultural corn fields, those containing only corn, continue to regenerate across the midwestern United States? An ultimate explanation that seems appropriate here appeals to the functions these ecological communities serve for the human industrial food system. It is because of these functions that we keep encouraging them to regenerate. This also applies in the microbial case. In medicine, current research focuses on how we can engineer a damaged human gut microbiome so that it continues to provide helpful functions for a human, such as digestion. Probiotic therapy (providing beneficial microbes), prebiotic therapy (providing molecular compounds that promote the growth of beneficial microbes), and fecal bacteriotherapy (the transplant of an entire microbial community) are all examples of microbiome engineering. As cases of gut regeneration become increasingly engineered, appealing to human design to explain why regeneration has occurred becomes increasingly appropriate.

As we explain in chapter 5, this engineering enterprise also importantly represents a different kind of scientific endeavor than the ecological and evolutionary ones discussed in earlier chapters; it is aimed not at knowing how and why regeneration works but at what we can do with this knowledge. How can we put this knowledge to the service of making pathological communities healthy again?

Beyond explaining how and why microbial community regeneration occurs, there are two further general ways that we see this book as helpful. We want to provide a challenging and new perspective on regeneration that is accessible to policymakers, biologists, historians, philosophers, teachers, and general readers, all of whom will benefit from understanding regeneration and imagining how the process carries across all scales of life. This perspective should also help readers recognize how all living systems are interconnected and impact each other. By recognizing these connections and applying knowledge of regeneration from one scale of living systems to others, we may be able to treat debilitating degenerative diseases and traumas and even heal our fractured planet.

Finally, in providing an introduction to microbial regeneration specifically, we hope to show how ideas and understanding gained from one living system can be applied to another. By looking at how regeneration works in different systems, we invite comparisons among them and provide a common language for researchers working on seemingly different problems and biological systems. This kind of cross talk has historically led to novel and exciting discoveries.[26]

Regeneration, in all its biological guises, has limits. Pushed beyond these, microbial communities do not regenerate the ecological functions they previously performed and that are so important for sustainability and medicine. Even in expert regenerators like *Hydra,* successful regeneration requires certain environmental conditions to be in place. Our hope is that the analysis of regeneration we offer in this book will provide some insight into sustaining regenerative abilities in the face of both natural and human-induced disturbance.

2 Ecology

> The problem with highly polarized issues is that they present only stark choices: is a community an organism or merely a collection of individuals that are laws unto themselves? The correct answer is "none of the above."
>
> DAVID A. PERRY, *Forest Ecosystems*, 1994[1]

Between the Blue Ridge Mountains and the flats of the Atlantic coastal plains is a hilly region in the southern United States called the Georgia piedmont (from French: "pied," or "foot," and "mont," "mountain"), covering approximately 30 percent of the state. Much of what was once old-growth forest dominated by stands of oak and hickory has been converted to agricultural land that supports the state's economy through crops, such as cotton and soybeans, and animal husbandry—the area's nickname is the "Poultry Capital of the World." These changes in the land have long intrigued ecologists interested in what would happen after an agricultural field is abandoned and humans cease to intervene directly upon the landscape. Given enough time, would an old-growth forest reemerge or would something else take its place?

Such questions ask about ecological community change over time and successfully answering them has been a central task of community ecology. These answers are also centrally important to understanding community regeneration, which is, after all, just one kind of community change: that resulting in the reemergence of what has come before, whether understood taxonomically (in terms of *who's there*) or functionally (in terms of *what they are doing*). To answer questions about community change, ecologists have sought the rules by which individual organisms from different species assemble into communities. These rules explain how community change occurs over time; they provide proximate explanations of community change. And they will help us answer our how-question about community regeneration. We will see, though, that answers to questions like that above have not only been centrally important but also the locus of controversy and heated debate. This debate, as we will also see, is relevant for our discussion of regeneration.

Those studying the Georgia piedmont, to continue our example above, have found that when an agricultural field is abandoned regeneration may occur, but it involves a gradual, century-long transition in the plant communities that live there. A community dominated by horseweed gives rise to one dominated by asters, which is, in turn, followed by communities dominated by broomsedge, then coniferous pines, and eventually something like the original oak and hickory mixed-deciduous hardwood forest (see fig. 1.2).[2] There are also corresponding transitions in the animal communities: for example, bird species change with the shift from grassland to pine forest to mixed-deciduous hardwoods, with each of those habitats offering very different lifestyle opportunities (table 2.1).

Long ago, ecologists noticed that regenerations, when they occur, are regionally specific. In the New Jersey piedmont, some 800 miles north of Georgia, the long transition that occurs after farmers have left their fields may also eventually result in an oak and hickory forest, but the transitional species are different—in sequence: ragweed, evening primrose, goldenrod, and junipers. Similar forests can regenerate by following very different paths. How can several paths lead to the same outcome? How predictable are these transitions in community composition? How are they affected by changing climates or pollution? These are the pressing questions surrounding community change in ecology.

Macrobial communities—those composed of large organisms visible to the naked eye—offer intuitively understandable cases of regeneration, but our focus here is on a much smaller scale. The microbial communities in our guts and on our teeth, for example, go through transitions in the diversity, abundance, and composition of microbes after they are disturbed by, say, antibiotics or brushing. Sometimes these transitions result in regeneration. Many of the same principles that apply to regeneration in the macrobial world also apply to the microbial, though there are some important differences that we get into below. So let's discuss how ecologists have thought about community change.

History offers resources for thinking through the issues, and it also helps set the stage for the contemporary ecological the-

TABLE 2.1. Number of mating pairs of common species of birds per 100 acres

SUCCESSION STAGE	GRASSES		GRASS-SHRUBS		PINE FOREST				OAK-HICKORY FOREST
Age in years	1	3	15	20	25	35	60	100	150+
Grasshopper sparrow	10	30	25						
Eastern meadowlark	5	10	15	2					
Field sparrow			35	48	25	8	3		
Yellowthroat			15	18					
Yellow-breasted chat			5	16					
Cardinal			5	4	9	10	14	20	23
Eastern Towhee			5	8	13	10	15	15	
Bachman's sparrow				8	6	4			
Prairie warbler				6	6				
White-eyed vireo				8		4	5		
Pine warbler					16	34	43	55	
Summer tanager					6	13	13	15	10
Carolina wren						4	5	20	10
Carolina chickadee						2	5	5	5
Blue-gray gnatcatcher						2	13		13
Brown-headed nuthatch							2	5	
Blue jay							3	10	5
Eastern wood pewee							10	1	3
Ruby-throated hummingbird							9	10	10
Tufted titmouse							6	10	15
Yellow-throated vireo							3	5	7
Hooded warbler							3	30	11
Red-eyed vireo							3	10	43
Hairy woodpecker							1	3	5
Downy woodpecker							1	2	5

TABLE 2.1. *Continued*

SUCCESSION STAGE	GRASSES		GRASS-SHRUBS		PINE FOREST				OAK-HICKORY FOREST
Age in years	1	3	15	20	25	35	60	100	150+
Crested flycatcher							1	10	6
Wood thrush							1	5	23
Yellow-billed cuckoo								1	9
Black and white warbler									8
Kentucky warbler									5
Acadian flycatcher									5

Data source: D. W. Johnston and E. P. Odum, "Breeding Bird Populations in Relation to Plant Succession on the Piedmont of Georgia," *Ecology* 37 (1956): 50–62.

Note: This table illustrates how the composition of bird species changes with corresponding changes in plant community type, as depicted in fig. 1.2: a grass community contains grasshopper sparrows and Eastern meadowlarks, whereas an oak-hickory community contains neither of these species but a host of others. The numbers corresponding to each bird species represent estimated pairs of birds per 100 acres (shading is added to make this information more accessible and apparent: the darker the shading, the higher the number).

ory we explore below. Debates about what constitutes an ecological community and the changes it can undergo date back more than a century. In 1916, Frederic Clements, whom we met in the last chapter, published a large and detailed manuscript summarizing an impressive mass of historical data about plant communities (fig. 2.1).[3] This was the first theory of community ecology, a novel and influential framework for understanding how and why communities change over time.

Clements was impressed by the fact that transitions from one kind of ecological community to another seemed to be surprisingly orderly and predictable. Working in the prairies of Nebraska, he closely studied the plant communities that fol-

FIGURE 2.1 | An undated photograph of Frederic Clements (1874–1945) with his wife Edith Clements (1874–1971) at the Carnegie Institute's Alpine Laboratory on Pikes Peak, Colorado. Frederic and Edith were inseparable, together except for one three-day stretch during forty-six years of marriage. See Joel B. Hagen, "Clementsian Ecologists: The Internal Dynamics of a Research School," *Osiris* 8 (1993): 184. (Edith S. and Frederic E. Clements Papers, accession 01678, box 69, folder 1; reproduced courtesy of the University of Wyoming, American Heritage Center.)

lowed one after another as old, abandoned wagon trails repopulated with plants. Defining his objects of study in terms of dominant species, he asked why does a community dominated by *little bluestem grass* (*Schizachyrium scoparium*) seem always to follow one dominated by *wickiup grass* (*Muhlenbergia pungens*), and why does a community dominated by *redwhisker clammyweed* (*Polanisia dodecandra*) follow the *little bluestem* community? Whence this order when we might expect chaos?

The order, which he called ecological succession, Clements answered, came from the fact that communities were a complicated kind of organism—a superorganism—and their stages were stages of development. The goal of succession, the ordered transitions he observed, was to arrive at a mature superorganism, what he called the climax community. That is to say, he believed the process was teleological: it could be explained by this end goal, as if that were its "purpose." The stages of succession, he wrote, "have the same essential relation to the final stable structure of the organism that seedling and growing plant have to the adult individual," and "just as the adult plant repeats its development, i.e., reproduces itself, whenever conditions permit, so also does the climax" community.[4] Just as the limbs of the axolotl regenerate according to a set limb pattern—you don't get an axolotl arm one time and a squid tentacle another—so too did ecological communities. He believed that as long as the climate remained the same, and humans did not interfere too much, the same community would regenerate, again and again.

Geography and climate defined communities. The oak and hickory forest of the Georgia piedmont was one superorganism, the pine forests of western Montana were another; each had their own unique sequence of successional development. Each sequence, however, followed the same general outline, just as individual organisms developed along predictable lines, say from a fertilized egg to a larva to a pupa to an adult.

Clements thought that succession had six "causes." It begins with *nudation*: caused by, say, forest fires or windstorms that denude a region of its plant species. Next, *migration* occurs as new plant seeds arrive from undisturbed areas nearby. If the environment is favorable, these new seeds begin to establish and grow, what he called *ecesis*. As these plants establish themselves, they change the physical and biotic elements of the environment—a cause he called *reaction*—setting up new niches that may favor the arrival of additional plants. As more plants arrive, *competition* between plants determines which species are able to establish themselves for the long term. These causes, Clements argued, could explain *how* succession occurred.

Nudation, migration, ecesis, reaction, and competition constitute five causes, but what about the sixth? It had to do with what was "directing" the process, so to speak, which Clements saw as providing an answer to *why* succession occurred rather than merely *how* it occurred. In other words, and using terminology from chapter 1, Clements saw this sixth cause as the key to an ultimate explanation of succession, and he contrasted it with the first five causes, which could be appealed to in a proximate explanation (an answer to *how* succession occurred). So, what was this sixth cause?

Even before Clements, ecologists knew that some combinations of plant species seemed to persist more or recur at greater frequency than others: such communities seemed to resist change; others were more fleeting or unstable. Why did some communities possess this stability? His answer was that *stabilization*, his sixth cause, was the "final cause," as philosophers put it, of the process and that these most stable communities were mature superorganisms; adult communities, so to speak. By invoking the notion of a final cause, Clements harkened back to the ancient Greek philosopher Aristotle who first theorized the idea. A final cause, á la Aristotle, is the purpose or function of something. The final cause of a dinner table is dining because that is the purpose for which the table is created.

If the language of causes seems odd here, and it is a little odd, think of it this way. What caused the dinner table to be built? The growth of the trees from which it was made, the workings of logging machines, and the skills of a carpenter (this is a bespoke table) all contribute to its making but, ultimately, what caused it to be built was the need for a dining surface. Clements argued that stabilization was the purpose of the successional process; the process was for stabilization, as a table is for dining. Changes in the composition of species in a community had a purpose: to create stable ecological communities.

If the claim that a stable ecological community is like a dining table sounds a bit strange to you, you're not alone. Even in his own day, Clements's theory was challenged, most notably by fellow ecologist Henry Gleason (1882–1975) (fig. 2.2).[5] Although Gleason initially gained just a few supporters, his critique is today widely embraced by ecologists as a welcome intervention. Gleason rejected Clements's appeal to final causes—succession, he argued, had no function or purpose. He believed that the patterns Clements sought to explain through superorganismal development could be better explained by proximate causes involving only the interactions of individual organisms with each other and with their abiotic environment.

Simply put, Gleason thought the world was not as Clements described it. Clements had made community change sound

FIGURE 2.2 | Henry Gleason (1882–1975) conducting research on tropical vegetation in Sri Lanka, then named British Ceylon, likely in 1914. Gleason took a yearlong leave of absence (1913–14) from the University of Michigan to travel the world. His goal was to gain a greater appreciation of the variety of different plant communities that exist on Earth. See Malcolm Nicholson, "Henry Allan Gleason and the Individualistic Hypothesis: The Structure of a Botanist's Career," *Botanical Review* 56 (1990): 91–161. (Courtesy of the LuEsther T. Mertz Library of The New York Botanical Garden.)

as predictable and orderly as organism development. He said communities could be easily distinguished from one another and classified, as one might distinguish two organisms from one another and classify them as members of different species. These ideas were central to his superorganism framework. The natural world, Gleason countered, was not so neat and, in fact, ecologists applying Clements's framework seemed to have a very hard time even agreeing on where a superorganism begins and ends. "We are treading upon rather dangerous ground," Gleason wrote:

> when we define [a community] as an area of uniform vegetation, or, in fact, when we attempt any definition of it. A community is frequently so heterogeneous as to lead observers to

> conflicting ideas as to its [. . .] identity, its boundaries may be so poorly marked that they can not be located with any degree of accuracy, its origin and disappearance may be so gradual that its time-boundaries can not be located; small fragments of associations with only a small proportion of their normal components of species are often observed; the duration of a community may be so short that it fails to show a period of equilibrium in its structure.[6]

And he continued with a specific example. Everyone knew that the beech-maple forest of northern Michigan, a favorite, indeed famous, site for early ecologists, was a single community. Yet, Gleason pointed out, "every detached area of it exhibits easily discoverable floristic peculiarities, and even adjacent square miles of a single area differ notably among themselves, not in the broader features, to be sure, but in the details of floristic composition which a simple statistical analysis brings out."[7]

Gleason's aim was not to undermine the idea of communities or community ecology. His intent was more subtle. He wanted to show that Clements's treatment of communities *as superorganisms* was just one *interpretation* of the data and there was room for alternative interpretations. According to Gleason, the weakness of Clements's interpretation was manifest in the confusion and disagreement among his contemporaries.

Gleason proposed a different way of looking at the problem. He introduced his alternative in the form of a rhetorical question: "Are we not justified in coming to the general conclusion, far removed from the prevailing opinion [that is, Clements's opinion], that an association is not an organism, [. . .] but merely a coincidence?"[8]

What did Gleason mean by coincidence? He did not mean that community formation was random and unpredictable. There was undeniably some order to the process: on newly emerged islands there can't be herbivores before there are plants, broadly defined. Coincidence meant simply the coinciding of a number of factors that give rise to changes in a community. Gleason was offering a different angle, one that focused only on the interactions among community members rather than the community itself. It was a bottom-up alternative, which rendered Clements's top-down approach unnecessary. According to the latter, everything that happened at the level of individual organisms could be explained by appealing to the stable, mature state of the superorganism. Clements invoked the goals of the higher level phenomenon, the superorganism, to explain lower level phenomena, and the interactions between organisms. Gleason argued instead that the interactions among individual organisms could explain the higher level phenomenon, ecological community change. His approach accounted for why ecologists could not agree about superorganisms: there were no such things. They were an unnecessary fiction. Whatever coincidental patterns occurred in the structure of communities, they were merely a result of lower level interactions—each species operating "selfishly," evolving under pressure to produce more progeny of its own kind but nothing more. There was no need to invoke the idea of a developing superorganism. And no need for teleology: succession had no goal.

By attending to coincidences in ecological interactions,

Gleason also emphasized the contingency of communities and that they could have been otherwise. Recall that Clements gave the impression that succession was contingent upon climate but otherwise nearly inevitable. Knowing the climate, one could predict what community would develop following denudation with near certainty. Gleason argued, in contrast, that community change was contingent upon more fine-grained factors, in particular the dispersal abilities of plants, their local environmental requirements, and chance events (like wind patterns, needed to disperse seeds in particular directions). Chance events produced different communities, thus the final structure of a community was much more sensitive to more—and more changeable—factors than just climate. This introduction of chance gave community change an unpredictability that was lacking in Clements's vision.

The story of Clements and Gleason is more relevant now than ever. Although this history is often told as a tale of two competing, even incompatible, visions for community ecology, it seems to us that their legacies (suitably modified and updated) are complementary, and both may be required to explain regeneration in different types of communities. After many years of abuse, Clements's ideas find resonance in current research challenging standard conceptions of organisms and their interactions, as we discuss in chapters 3 and 4. But, for now, we shall largely build on Gleason's ideas about the role of coincidence in the formation of ecological communities.

A century on, theory in community ecology is a tangle of different principles, an "intangible mess" to some.[9] One way to

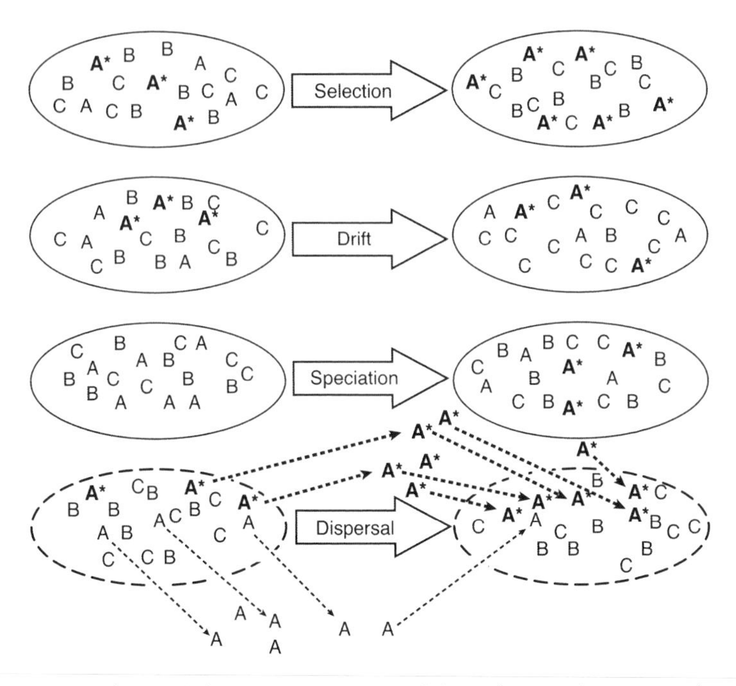

FIGURE 2.3 | Vellend's four processes in ecology. In "selection," species A* outcompetes (is fitter than) species A and comes to dominate a community over time. In "drift," random sorting in small samples by chance replaces most individuals of species B with individuals of species C. In "speciation," species A splits into two species, A and A*. In "dispersal," individuals of species A* come to dominate over individuals of species A, simply because they are more readily recruited into the ecosystem.

gain traction and give order to this tangle is to follow community ecologist Mark Vellend, who describes community change as resulting from four processes: selection, drift, speciation, and dispersal (fig. 2.3). He summarizes:

> at the most general level, patterns in the composition and diversity of species—the subject matter of community ecology—are influenced by only four classes of process: selection, drift, speciation, and dispersal. Selection represents deterministic fitness differences among species, drift represents stochastic

changes in species abundance, speciation creates new species, and dispersal is the movement of organisms across space.[10]

If you have taken an introductory course on evolution, you may recognize the names of Vellend's processes. He likens these four processes in ecology to the four causal factors biologists working in the field of population genetics use to understand evolution (selection, drift, mutation, and migration). This comparison helps to clarify "ecological assembly rules," the phrase modern ecologists use to describe the rules that determine the formation of an ecological community. We will describe his framework here, but because we, unlike Vellend, are primarily interested in microbes, we will also have to modify it slightly.

Consider an example from the chapter 1. The human gut harbors a complex community of microorganisms that provides many helpful services for its host.[11] Imagine that you have just taken broad-spectrum antibiotics. Even if they were not the intended target of treatment, the microbes in your gut will also be affected and many will die; this is why some people find that antibiotics are "hard on the stomach." You'd like to get back to normal, to restore the ecological services provided by the community living in your gut. What ecological processes might affect its regeneration? What do you need to know about in order to know whether regeneration will occur?[12]

Well, processes of *dispersal* matter. The composition of the community in your gut will be affected by what species arrive in the area. Who is where and when influences community dynamics. A species that doesn't arrive cannot join the community (think: if there are no cardinals in your part of the world, cardinals will not join the birds at your feeder, regardless of how perfect for cardinals the food you provide is). A species that arrives early may change the environment and make it suitable for some, but not other, species. Species that arrive in abundance have a better chance of persisting. This is Gleason's idea about the importance of migration.

For a long time, it was assumed that although dispersal matters greatly for the macrobial world, like cardinals, it didn't matter as much for microbes. This is captured in the microbiologist's saying that, when it comes to microbes, "everything is everywhere, but the environment selects."[13] But current philosophers and microbiologists are challenging this assumption. One example is the work of Martin Blaser, an infectious disease expert, who argued in his 2014 book *Missing Microbes* that our wide use of antibiotics has significantly altered the microbes in our environment, affecting whether regeneration can occur for us as it did for our ancestors.[14] We will return to this in chapter 5, but the point now is that dispersal matters for community dynamics, even for microbial communities.

Vellend's second process, "speciation," might have only a long-term effect on communities of macrobes because speciation of such organisms is understood to be a slow process. But something like it might well speed up the evolutionary process in microbial communities and affect their short-term regeneration.

In macrobial ecology, to start with the kind of communities Vellend primarily has in mind, speciation can occur through character or trait displacement: one species becomes two as its members exploit resources for which there is less competition. Over time the two strategies lead to distinct characteristics and eventually distinct species.[15] Charles Darwin's famous finches are an example. Darwin witnessed character displacement among the finches of the Galapagos Islands, weaving it into his theory of evolution. He surmised that differently shaped beaks evolved as finches migrated among islands dominated by different types of plants. When a species of finch arrived on an island characterized by plants with several differently shaped seeds, over time some finches evolved different beak shapes and sizes to better consume the novel seeds. There would be less competition for food that way. Given enough time, these changes led to the creation of two species out of one, each exploiting different seed types. Importantly for community ecology, this speciation of finches didn't just increase the number of species of finches in the Galapagos; it also increased the number of traits within a community of birds on any one island: the community now contains species doing different things. So, speciation might, in the very long term, have an influence on the structure of macrobial communities (both taxonomically and functionally, to use our terminology from chapter 1).

One genetic mechanism that might speed up an analogous speciation process for microbes would be the acquisition of genes by lateral gene transfer. Lateral gene transfer, rare in "macrobes" but common and important in the microbial world, occurs when genetic material is directly transferred from one bacterium to another (see fig. 4.1).[16] The concept of species is problematic when applied to bacteria, for the reasons discussed in chapter 1, but the diversification of bacteria via lateral gene transfer is well documented and is often stimulated by living together in a community, ultimately resulting in more diverse communities. In other words, "speciation" might be an awkward phrase when used to discuss organisms for which species is a questionable category, but lateral gene transfer provides an analogous process that increases diversity. (So here we use Vellend's "speciation" terminology broadly to include lateral gene transfer, so that our text remains consistent with his theory.)

In macrobial organisms like Darwin's finches, speciation takes a long time, but the acquisition of new genes by lateral gene transfer makes the process of diversification an important short-term factor influencing change in microbial communities. And it is for this reason, too, that many experts caution against overusing antibiotics: microbes can evolve quickly to evade them.[17] Many of the antibiotic-resistant "superbugs" increasingly endangering human health owe their resistances to lateral gene transfer. Just imagine if macrobial organisms could do this. What if a salamander could pass on the genes required to regenerate limbs to humans, and this activity was further enhanced when the humans were at war and more likely to need the help! Lateral gene transfer enables microbes to evolve widespread antibiotic resistance. If one community member happens to be resistant to a particular antibiotic it can pass this trait directly to other members.

Returning to the human gut microbiome, and pulling together what has been said so far, regeneration can occur as a result of dispersal and speciation. Community function lost when the system was flushed by antibiotics may be restored through the arrival of new species that can perform the required functions (dispersal), or the species that survived the antibiotics may evolve by lateral gene transfer to perform different tasks in the gut, replacing the work done by lost species (speciation).

Moving beyond speciation and dispersal, Vellend uses *drift* in pretty much the same way as population geneticists and this requires some discussion. In genetics, drift is the name for stochastic ("chancy") influences effecting changes in gene frequencies in a population, more significant in small populations than large (this is sometimes called "sampling error": a sample of a few individuals is more likely than is a sample of many individuals to be skewed by chance). Since no real-world population or community is infinitely large, chance will always play a role in both genetics and community change. A simple example from population genetics: a certain deer may be more fleet than those in the rest of the herd, and thus more likely to pass on its genes, but a chance event, like the deer being hit and killed by lightning, will wipe out its genes regardless of their fitness. This is a stochastic effect but one influencing gene frequencies in the population of deer.

In ecology, Gleason is often considered the first person to emphasize the importance of chance events that can affect a community's species composition—what is now called "ecological drift." His example was wind patterns, which might add an element of chance to the arrival of seeds, and thus to who can join the community. All processes which affect species composition in stochastic ways Vellend considers to be causes of ecological drift.

The fourth class of process is "selection." Ecologists like Vellend use "selection" in a specific ecological way: to refer to ecological processes that discriminately sample (or "choose," so to speak) members of a community. For them, selection occurs when organisms within species A outcompete (have more progeny than) organisms within species B in a population comprising all A and B organisms (and possibly more) because of some phenotypic property organisms of species A share and organisms of species B lack. That organisms of species A are selected on the basis of certain properties they possess is what makes this a *discriminate* sampling process (as opposed to ecological drift, which is an indiscriminate process).

This is different from the way that the term "selection" is used in evolutionary biology. For evolutionary biologists, selection describes only discriminate sampling of organisms *within* a single species (A or B, as above) population. It is the frequencies of individual organisms of a specific type *within a single species* (say A) that selection affects, and organisms of different species (say B) with whom individual members of A may indeed be competing (or eating) are to be taken as part of the environment. These differences can lead to confusion, but highlighting the differences seemed to us preferable to covering them up by adopting some language used by neither group of biologists.

Throughout the twentieth century, ecologists have often

treated competition for resources between organisms of different species as the main process through which species were "selected" to be members of a community. Competition eliminates certain members of the community (and their species with them) that cannot keep up, thus "selecting" those that are better competitors. But competition is only one kind of interaction relevant here. Predation, for example, can select against members of the predated species (at least in the short term). And mutualistic interactions, cooperative behaviors that are beneficial to all or most parties involved, may increase the relative proportion of the species exploiting them. One example is the production of so-called public goods in microbial communities: some bacteria release compounds into the extracellular environment that provide a collective benefit, such as a widely required nutrient. Each of these processes (competition, predation, mutualism) is said to be "selective" insofar as it discriminately "chooses" among species *within* a community.

These examples of selective processes all have to do with interactions *among* members of the community, but the background environment also acts as a selective sieve. To see how this is the case, consider again our antibiotics example: we can divide the selection processes that determine community composition into two types. First is the local environment of the gut, which is in part controlled, of course, by the host human. This environment discriminately selects certain microbes that can live on the mucus layers produced by the lining of our guts, and that can evade being targeted by antimicrobial compounds regulated by the immune system. These are akin to Gleason's "environmental requirements": the phenotypes (physical and biochemical characteristics) of certain microbes dispose them to live well in the gut environment (and, of course, in many cases, they've evolved these phenotypes for this reason). The second type of selection process involves the interactions *among* the microbes themselves, such as competition for resources or predation.

Importantly, especially for our purposes later in this book, there is an element in common between Vellend's use of selection and that term as used by evolutionary biologists. As explained in chapter 3, *no selection on communities as systems favoring their (the community's own) survival* is implied in either case.

Let us summarize. These four classes of process together determine community dynamics—that is, how a community will change through time and how it will respond to disturbance—and thus they determine ecological regeneration, which is, after all, just a kind of community change. How your gut will respond following disturbance by antibiotics depends on the nature of the disturbance and how these various processes operate. Strong environmental selection (by the host's immune system, say), might direct regeneration toward a community beneficial to the host. Limited dispersal of microbes will put a limit on the kind of community that can regenerate—a limit that might be overcome by the process of speciation or lateral gene transfer. If the disturbance severely damages the community, leaving only a few individuals, then drift may initially (and potentially in the long term) affect whether regeneration occurs. These are the same processes that affect free-living microbial communities (that is, microbial communities not associated with hosts), as well as macrobial communities, like the Georgia piedmont. And, of course, in all real communities, most or all of these processes are operating simultaneously.

Many of the most heated debates within twentieth-century community ecology concerned the relative significance of these four processes in determining community change. Is dispersal more significant than competitive interactions? Does ecological selection outweigh ecological drift? Is speciation ever significant, or is it too slow a process to matter on the timescale of community formation? As ecologists realized that answers to these questions depend on the particular community under consideration, these debates have become more localized, and now the question is which kinds of processes matter for any particular community, or which matter for explaining differences between different communities.

The important theoretical point is this: Vellend's four classes of process together capture all the proximate causes of community change through time and thus dictate whether community regeneration will occur. Of course, how and whether regeneration occurs in any specific case will depend on the details of that case (as we discuss further in chapter 5). But abstracting from those details, we now have an understanding of all the kinds of proximate causes that direct community regeneration. To summarize, regeneration is, proximately speaking, partial or complete community assembly according to the already evolved properties of organisms within species. How does regeneration occur? In any specific case, it occurs through the action of these processes. As Mayr, quoted in chapter 1, might put it, if regeneration occurs, we know that these processes will either separately or in concert comprise "the pathway by which it has been achieved."

It is worth emphasizing, again, and to foreshadow our discussion in chapter 3, that this does not require that communities themselves (either macrobial or microbial) be subject to natural selection *as communities*. Community evolutionary trajectory and stability could be just a fortuitous by-product of natural selection impinging on the species that make up the community. Insofar as species have been evolved in a community context, they will have been "coevolved." Coevolution will be an important concept in the remainder of this book, and it simply means that species A and B have each acquired adaptations to the presence or activities of the other, as if they were each an important part of the other's "environment": B is part of the environment of A, and A is part of the environment of B. There is no necessary evolution of traits that might promote the survival or replication of the *community* (A + B) itself. It could be that individuals of A have coevolved with individuals of B in such a way as to eat or destroy them (an example of predation), and any community dependent on B individuals that acquires an A individual is doomed as a community. Coevolution, in other words, need not imply cooperation and need not be benign. Already SARS-CoV-2, known as COVID-19, is coevolving with *Homo sapiens*, the better to infect and spread within our populations, while humans with innate resistance are (presumably) increasing slowly in numbers. Attenuation (reduction in the severity of infections) or, alternatively, an increase in severity could be the result in the next few years. It all depends on population dynamics, selection, and chance.

It is one thing to provide a unified theoretical framework for understanding the ecological interactions that influence community change and quite another to predict, or in the case of medicine and sustainability, engineer, the direction of change of any particular community. With a discussion of history and contemporary theory at our disposal, we end this chapter by revisiting the proximate-ultimate distinction.

Clements, recall, offered both proximate and ultimate explanations for regeneration. His proximate explanation had to do with the first five of his six causes. But Clements also thought regeneration performed a function for the community—it helped the "superorganism," as he called it, respond to damage and maintain itself. That is, he also thought regeneration occurred because of the function it performed. Gleason, on the contrary, offered a detailed proximate explanation and thought that Clements's ultimate explanation was unsupported by data and, anyway, superfluous. Regeneration could be fully explained proximately—no need (or justification) for functions or purposes. Our discussion of current theory shows how modern ecologists have built on Gleason's vision. There may not seem to be much room left for Clements's vision or for ultimate explanations of regeneration based on evolution by natural selection, such as those discussed in our first chapter. Vellend's account suggests what processes have led to "more successful" rather than "less successful" multispecies communities but give us no reason to believe that community success itself has anything to do with the theory of evolution by natural selection.

Twenty years ago, this could have been the end of the story: community regeneration has a proximate explanation but no ultimate explanation. But the rise of the science of metagenomics, and the increased depth it has brought for our understanding of microbes and their importance for animal and plant life, has had a disrupting (in a good sense) and exciting effect: it is motivating, once again, the search for ultimate explanations of community regeneration. The current biological literature is now replete with attempts to make sense of communities as entities that evolve by natural selection in their own right. This is in part because of the revival of something like Clements's superorganism view, albeit in a modified guise. Unlike Clements, who claimed that communities were organisms, this new work argues instead that organisms, humans included, are actually communities with organism-like properties! Communities made up of macrobial "hosts," like plants and animals, and their associated microbial communities, like the microbes on our teeth or in the bee's gut can, in such a view, evolve by natural selection. "We have never been individuals," as three proponents of the view put it.[18] These newly described biological entities are often called holobionts. What is this new research and how does it square with what we've written about ecology in this chapter? In order to answer these questions and assess this new view and those related to it, however, we will have to move from ecological to evolutionary thinking.

3 Evolution

Evolution by natural selection is change in a population owing to variation, heredity and differential reproductive success.
PETER GODFREY-SMITH, *Darwinism and Cultural Change*, 2012[1]

Biologists frequently employ the language of purpose and function when explaining the natural world. They even do so with a casualness that is perhaps unique among scientists. What is the *purpose* of the hummingbird's long, needle-like beak? What is the *function* of the human nose? This language is rendered sensible through evolutionary theory; this theory, and this language, will be our subjects now. Specifically, in this chapter we take stock of traditional Darwinian thinking about community evolution and regeneration. Our aim is to assess whether the restoration of a microbial community after some disturbance (what we've called community regeneration) and the regrowth of a part of an organism after some catastrophic event (what we've called organismal regeneration) are really in some fundamental way similar. The way forward will involve thinking through their evolutionary functions or purposes.

We define the traditional Darwinian way of thinking about evolution as a combination of the three-part formula of the famous Harvard evolutionary biologist Richard Lewontin (often called "Lewontin's Recipe") and multilevel selection theory, both of which we explain in this chapter. We argue that this combination does not quite do the job, at least not as usually put together, even by those who would be most sympathetic to our project. We'll use as examples of this *not-quite-doing-the-job* the modestly sized microbial communities associated with animal and plant hosts (the ensembles sometimes called "holobionts") and the largest largely microbial community there is, the biosphere. In chapter 4, we suggest ways around this predicament, using an alternative framing of *natural selection*. Within such a framing, it does make sense to think about the regeneration of limbs and of communities as being *ultimately* similar.

Let us begin by returning to the relation between organismal and community regeneration as introduced in chapter 1. We are now in a better position to discuss their respective explanations and to explore what is the same or different about them. The mechanisms or procedures by which organisms regenerate limbs and by which forest landscapes "regenerate" something like their previous "natural" states after a fire or a period of human use—or our gut microbiota "recover" after a course of antibiotics—are clearly different. Organismal regeneration is, *proximately*, a process involving communities of interacting cells identical or very similar in genotype and the turning on and off of the expression of various genes. A pro-

TABLE 3.1. Proximate and ultimate causation/explanation in organismal and community regeneration

	PROXIMATE	ULTIMATE
COMMUNITIES	Ecological assembly	Natural selection as differential recurrence/persistence
ORGANISMS	Developmental genetics	Natural selection as differential reproduction/replication

Note: See text of chapters 3 and 4 for details.

cess resembling, at least in its outcome, development during embryogenesis of the structure now missing must occur. Something resembling a lost structure or at least having the same function must be produced.

On the other hand, community regeneration is (again, *proximately*) a matter of already coevolved ecological interactions between individuals of different species and often very different genotypes, as reviewed with examples from macrobial and microbial communities in the chapter 2. The several species involved in the formation of a community have already established and evolutionarily fixed properties and propensities, and it's largely on the basis of these that communities form, following ecological community assembly "rules." There can only be herbivores where there are plants, for instance. Of course, genes are involved in both "regenerative" processes, and lateral gene transfer (chapter 4) may play a role in both, but the languages of molecular and developmental biology are used to explain the former, while the language of ecology governs the latter, *proximately* speaking (table 3.1). Different languages, different concepts, different sorts of proximate explanation, with the differences bringing us back to the larger question about regeneration—whether it is *ever* really the same kind of thing across different scales of life.

If there is a unified way of explaining regeneration across scales—that is, of seeing organismal and community regeneration (either for macrobes or microbes) as effects of the same or even similar causes—then it must be the *ultimate* cause that is the same. Different "hows" but, maybe, the same "why." Recall that ultimate causes explain what a trait (regeneration in this case) is *for*, its *function* or *purpose* in other words. Following Ernst Mayr, the leading evolutionary biologist in the middle of the last century, many of us now conceive of ultimate causes in terms of the theory of evolution by natural selection, and it's to that theory that we will turn.

Generally (with more detail later), standard current formulations employ Lewontin's Recipe, summarized in his words as follows.

> A sufficient mechanism for evolution by natural selection is contained in three propositions:
>
> (i) There is variation in morphological, physiological, and behavioral traits among members of a species (the principle of variation).
>
> (ii) The variation is in part heritable, so that individuals resemble their relations more than they resemble unrelated individuals and, in particular, offspring resemble their parents (the principle of heredity).

> (iii) Different variants leave different numbers of offspring either in immediate or remote generations (the principle of differential fitness).
>
> All three conditions are necessary as well as sufficient conditions for evolution by natural selection.[2]

That is to say, there will be evolution by natural selection in any situation in which there is *heritable variation in fitness*. This so-called recipe is like the instructions for baking bread: when these ingredients come together in the right way, natural selection automatically results. It's in this context that we ask, "If organismal regeneration is to be explained by evolution by natural selection targeted to organisms within species, can we understand community regeneration as the result of selection *targeted to communities* within the biosphere as a whole?" And we will conclude that with standard formulations of the theory of evolution by natural selection we cannot, but by tweaking them we can.

We should emphasize that it's not necessary that there be any ultimate or evolutionarily adaptive cause for regeneration to occur at either level. As discussed, it could be that regeneration of a limb involves just the turning on and off of the same genes as involved in normal embryonic limb development. Such proximately caused regeneration could then be viewed as a more or less automatic or even accidental response, a replay of the developmental programs involved in the initial formation of limbs, triggered by the simple fact that one is missing. It's only if organisms that regenerate limbs generally have more progeny than those that don't and this regenerative ability has been selected for and honed to its current state of reliability over many previous organismal generations—precisely because of that reproductive advantage—that we can say that there is an *ultimate* cause involving natural selection. In such cases, regeneration was selected for, and survival through intact limb use in order to reproduce is its purpose or function. Evolution by natural selection seems to make no sense otherwise: if there are not reproducing entities that have more progeny as a result of their possession of certain heritable, fitness-enhancing, properties, there's no selection. (Of course, proximate mechanisms might well include a mix of processes involved in initial limb formation and processes specific to and selected for limb regeneration as an ultimate goal: often there is such a mix.)

Similarly at the community (forest or gut) level, we might say that community regeneration is "automatic" or "accidental" and needs no ultimate cause, if it is taken to be only a fortuitous by-product of previously coevolved relationships between the organisms or species involved. Yes, such collective behavior would result from *selected-for* responses, but these were selected for—were ultimately caused—at the levels of the individual species involved, not of the communities *as* collectives. Oaks and hickories follow conifers in the succession observed in the Georgia piedmont because the former have evolved to use soils "prepared" by the latter, not because forest succession is itself under selection or because regeneration could be seen as its "purpose" in the sense that Clements invoked as his sixth "cause." This was Gleason's point.

Similarly, restoration of a healthy gut microbial community could simply be a result of recruitment of species that can survive there. Species or the individuals within communities comprising them might be said to exhibit purposes evolved by natural selection, but, again, the communities they make up need not. *Coevolution* in the sense defined in chapter 2 is not necessarily favorable to any community-level function or stability. In this view, coevolved communities are assembled by ecological principles as in chapter 2, but such assembly processes are not themselves subject to natural selection the "purpose" of which is the perpetuation of properly or beneficially assembled communities.

Indeed, there's already a highly elaborated theory involving coevolution that could do this proximate work. Each species, through activities selected for because they result in the differential reproduction of individuals within the species, inevitably modifies the environment in which it lives. Inevitably, this modifies the conditions under which individuals in this, *and other,* species evolve (again, by differential reproduction). Any benefits at the community level are just "fortuitous by-products." The theory here is *niche construction* theory, and it offers an easy way of explaining proximal causation as the result of coevolution without invoking community-level selection or community-level adaptation.[3] A very popular example as an alternative to the Georgia piedmont or the human gut would involve beavers, whose dam-building (pond-creating) activities modify the conditions under which pond-dwelling fish and water-insect species then evolve. There need be no selection on the beaver-fish-insect community as a whole.

If there really is nothing common between organismal regeneration and the "regeneration" of communities, proximately, other than that both do look like a return to "normal," and there is nothing ultimately the same either, we could end this book now. Why thinking this way would be the default, and why thus traditional evolutionary theory fails us in the cases of many multispecies collectives, is the subject of the rest of this chapter. Both are central to our argument. However, we would like to take a more positive approach, asking how we might rework evolutionary theory so that we can talk in terms of natural selection and *ultimate* causes of the regenerative behaviors of communities as well as of organisms. This is the subject of chapter 4.

To clarify our point further, it's worth saying that some have argued that microbial communities are intermediate, needing both proximate (ecological) and ultimate (evolutionary) explanatory paradigms. After all, real-time evolutionary changes often do occur in the life of a microbiome, even that associated with a single host.[4] If so, then selected-for processes, such as the transfer of antibiotic resistance genes by lateral gene transfer between species, might occur in the evolution of the gut microbiome of a single human over times much shorter than his or her lifespan. As well as the human gut being a site for "automatic" ecological interactions between species, reflecting thousands or millions of years of prior coevolution of gut microbes with each other and us, some real-time natural selection could be happening, too, at the species level.

But, again, this is within-species selection; our concern is that standard evolutionary theory cannot deal with selection for *collective* processes that affect the regenerative ability of multispecies collectives (communities) as wholes, because such collectives do not reproduce as wholes. After reviewing in more detail what the standard theory is, we show why it is inapplicable to most cases of microbial communities, both at a very small scale (the case of "holobionts," collective entities comprising a multicellular animal or plant "host" and its associated microbial communities) and at the largest scale (the Earth's whole biosphere, the biotic component of "Gaia"). All this detail may make the rest of this chapter a hard go, but it's at this level only that we can see why naïve invocations of vague selection processes affecting holobionts or the biosphere are wrong and wrongheaded.

As we've said above, standard formulations of natural selection are usually underwritten by Lewontin's Recipe. Richard Lewontin, one of evolutionary biology's most influential figures, regarded his three conditions (variation, heredity, and fitness) as both necessary and sufficient for evolution by natural selection, and many have since accepted this. But neither heredity nor fitness makes any sense except in the context of reproduction: selected entities, according to Lewontin's Recipe, are those that differentially reproduce.

Peter Godfrey-Smith, a philosopher and articulate advocate of traditional Darwinian thinking currently, endorses and updates Lewontin's view, writing that,

> Evolution by natural selection is change in a population owing to *variation, heredity* and *differential reproductive success* . . . the

> criteria required are abstract; genes, cells, social groups and species can all, in principle, enter into change of this kind. For any objects to be *units of selection* in this sense, however, they must be connected by parent–offspring relations; they must have the *capacity to reproduce*. Units of selection in this sense can be called *Darwinian individuals*.[5]

The updating here, to which we will often return, more fully recognizes (as many theorists like Peter Godfrey-Smith, and we ourselves, now do) the legitimacy of what is called multilevel selection theory. According to this theory, Lewontin's Recipe is not limited to organisms (members of a species); note that Godfrey-Smith intentionally refers only to "change in a population" rather than a population *of organisms*. If something like reproduction (or the making of reasonably similar copies) can be said to occur among entities at any level of the biological hierarchy—usually taken to include the nested levels of genes, cells, multicellular organisms, circumscribable groups of these (beehives, for instance) and species, then they *too* are subject to natural selection and evolve in part through its agency (table 3.2).

At the bottom of the biological hierarchy are genes (stretches of DNA). Selfish DNA theory, as it is called, recognizes that stretches of DNA can make extra copies of themselves (be differentially reproduced—replicatively amplified by duplication or transposition) and are thus selectable entities.[6] Transposition describes the several mechanisms by which a specific region of a chromosome (a "transposable element") deposits itself or a copy of itself in some new chromosomal position.

TABLE 3.2. The biological hierarchy

UNIT	EXAMPLE	MEANS OF REPRODUCTION
Biosphere as a whole	"Gaia"	None
Ecosystem	Forest	None
Multispecies community	Tooth biofilm	None
Species	Homo sapiens	Speciation
Single species community	Beehive	Specialized members (queens and drones)
Multicellular organism	You or me	Sex
Single-cell organism	Bacterium	Binary fission
Gene	Gene for antibiotic resistance	DNA replication

Note: The so-called biological hierarchy, a sort of Russian-dolls-type nesting of individuals into more inclusive individuals. Only entities at levels in which there is a means of reproduction can evolve by natural selection as such is traditionally understood.

About half of our own (human) genome comprises such virus-like elements, and whatever organism-level "function" they might sometimes now perform, there is general agreement that such elements arose through selection at the gene level. In other words, stretches of DNA capable of using the cell's DNA replication activities to make more copies of themselves were selected to do so, regardless of consequences for the cell, as long as these extra copies were not too deleterious. Cells in multicellular organisms (again, we humans are an example) are subject to selection too, occasionally escaping organism-level controls on their replication and thus causing cancers: cellular selection models are well developed in oncology. Selection on social groups often leads one to dominate another: controversies over whether or not altruism (individuals sacrificing their own reproductive interests for "the good of the group") leads to differential group survival still rage, but there is no doubt that this is possible under some conditions. Species selection, resulting from the differential diversification (meaning: speciation rate in a genus exceeding extinction rate in that genus) has been accepted by many paleontologists for several decades now as an explanation for the prevalence of certain groups in the fossil record.[7] Frequent speciation will in general increase the number of species.

Again, the *only* requirements for evolution, according to multilevel selection advocates like Godfrey-Smith, are *heritable variation* in *fitness* (Lewontin's Recipe). Heredity is here defined simply as parent-offspring resemblance in properties appropriately defined for the relevant level of the hierarchy, however reproduction or replication happens, from DNA synthesis through to species formation.

Such current multilevel views of evolution by natural selection, especially as they pertain to levels above the organismal, might be contrasted to the gene centrism of the influential theorist George Williams and his popularist Richard Dawkins, evolutionary biologists whose thinking dominated the field for much of the last century. Both insisted that one should not infer selection at any level of the biological hierarchy higher than necessary. And this position is still supported by some more conservative theorists.[8] These theorists would point out first that all higher-level (for instance species-level) traits like genetic variability can be attributed to lower-level (individual organism) causes like recombination (reshuffling at the DNA level), and, second, that higher-level populations are most often much smaller than the lower-level ones that of necessity make them up and thus more likely to be at the mercy of drift, not selection. In the first case, a sort of tough-minded reductionism is at play and, in the second, a possibly inappropriate extrapolation of the notion of a population and its vulnerability to "drift." Drift, as noted in chapter 2 is sometimes seen as a sort of sampling error. We are more likely to get around 50 percent heads if we flip a coin a thousand times than if we flip it just a few times, and there are usually many more organisms in a species than there are species in a genus, so chance would have a larger role to play if that's how we extrapolate the notion of population to the next higher level.

But this gene-centric reductionism seems to us and many advocates of multilevel selection to be unjustified. Our take is that higher levels have selectable properties that lower levels by definition cannot have, even if properties of organisms underwrite them. For example, how many individual organisms there are in a species, or how variable its members are in phenotype or genotype, are simply not traits that can be said to be a property of any individual organism within that species. Yet surely these are measurable collective properties of species as wholes and impact species diversification rate (speciation minus extinction). Diversification *is* subject to evolution by natural selection at the species level. To get more specific, one popular explanation for the prevalence of sexual species is that asexual species which might be derived from them lose the ability to reshuffle their genes (no mating and no recombination), and thus are less capable of meeting whatever challenges a changing environment might throw at them in the future. Thus asexual species are more prone to "go extinct." The species-level property here is variability of the gene pool, and this is simply not a property of any individual organism within a species of either kind, even though it may depend on many genes determining sexual behavior at the organismal level.[9]

Our take on the second objection to multilevel selection theory is that, although there may well often be fewer species in a genus than organisms in a species, there's no reason to assume that a genus or any other taxonomic ranking is the relevant population in species selection rather than, say, all contemporaneous species in a broadly defined ecological niche no matter how they are related. Populations are hard enough to define even for sexual species, which, according to Ernst Mayr's hegemonic *Biological Species Concept,* are interbreeding natural populations that are reproductively isolated from other such groups. Indeed, frequency of interbreeding as now assessed from DNA sequencing is one way to define the boundaries of a species. But how are we to define populations for groups that (by Mayr's own definition) do not interbreed, that is, for genera and higher—anything more inclusive than sexual species? If species selection is the answer to "Why are there so many species on Earth that reproduce sexually?" then the relevant population might as well be all species on Earth, now estimated to be in the millions.[10]

But even with our admittedly very generous attitude toward multilevel selection—that at any level at which we find entities showing heritable variation in fitness natural selection *must* occur—there are two substantial challenges associated with bringing the theory of evolution by natural selection to bear on the regeneration of multispecies communities. It turns out that the currently accepted formulation of that theory cannot be applied to multispecies communities in the way that it can be applied to genes, cells, organisms, social groups, or species. So more is needed to defend the idea that community regeneration has an ultimate explanation.

The first problem we must solve is that communities generally do not "reproduce" (directly make even imperfect copies of themselves) as systems, so they cannot evolve by natural selection according to Lewontin's Recipe. Multilevel formulations like that of Godfrey-Smith or ours (table 3.2) still do require at least this. We can call this first problem NO-REPRODUCTION. In this chapter we will focus almost exclusively on this problem.

The second problem, discussed in detail in the next chapter, is that what counts as community regeneration is often "functional" (requiring only that metabolic reactions or patterns of species interaction comparable to those of the initial state be recovered) rather than "taxonomic" (requiring that the same species are involved in implementing them). To use a metaphor, it is the "song" performed, rather than the "singers" performing it, that regenerates. Only this second problem would have an analog at the organismal level, which would be the involvement of different genes or different regulatory responses of the same genes during regeneration of a structure *versus* during its initial development. Lens regeneration, common in some amphibians, provides many examples, as does limb regeneration in axolotls or, as mentioned previously in chapter 1, spinal neuron regeneration in the lamprey.[11] We can call this second problem SONG-NOT-SINGERS.

Both problems are brought to the fore by debates over

selection as it impinges on microbial communities, and here we discuss two sorts of communities over which such debates have raged. The first, at a small (physical) scale comprises many "holobionts," which are collections of microbial species together with the hosts that they most often live on or in. The second, "the microbial engines that drive Earth's biogeochemical cycles," as Paul Falkowski and colleagues put it, encompasses the global (and largely microbial) community generally considered to be the biotic component of "Gaia."[12] Often, the Gaia hypothesis is taken as key to the development of a very active discipline necessary to human survival, Earth system science. But viewing the Earth as an organism-like "system"—if that is taken to mean that the Earth is an integrated entity with self-sustaining functions (for instance biogeochemical cycles) arising through evolution by natural selection at the "system" level—encounters these same two problems (NO-REPRODUCTION and SONG-NOT-SINGERS). Earth does not reproduce, and the "song" it sings is, over time, produced by different "singers."

We begin with the first type of community. The term "holobiont" was supposedly first coined thirty years ago by biologist and popularizer Lynn Margulis, but the "hologenome theory of evolution" was formulated and popularized much more recently, no doubt reflecting the increased ability to collect, and enthusiasm for collecting, "metagenomic" data from all the microbiota in and on "macrobial" hosts, humans in particular.[13] A "hologenome" is the sum of all of the DNA of a host and of the microbes associated with it—all the genes of a holobiont, in other words.

The metagenomics approach itself emerged out of frustration and opportunity in the early decades of this century. The frustration was in the slowness of culturing individual bacterial species or strains from the wild (or the gut): indeed some species cannot be cultured at all in the laboratory. The opportunity came from the astonishing drop in the costs and effort involved in sequencing all of the DNA (or all of the RNA or all of the proteins) in any sample, whether it be a scraping from a showerhead, a cupful of pond scum, or an individual human feces. Sometimes metagenomic data can be reassembled into genomes representing some of the predominant species present in the sample, but sometimes it's just the genes for relevant metabolic activities that are analyzed. Large metagenomic databases are now available, some devoted to cataloguing sequence data from the hundreds to thousands of microbial species characteristically in or on any animal or plant "host." Since many host properties do depend on associated microbes, or develop in conjunction with such microbes, notions of biological individuality and evolutionary trajectory often address the "holobiont" as a whole.

How does the NO-REPRODUCTION problem apply in the case of holobionts? Several early versions of the hologenome theory included claims that holobionts are "units of selection."[14] But this only makes sense according to Lewontin's Recipe or multilevel selection theory in those cases in which a host's microbes are guaranteed by a relevant coreproduction mechanism to be the parents of the microbes of that same host's progeny. In such cases, there is in effect just a single (albeit composite) reproducing unit, and this gives rise to definable "parent-offspring relations" for this holobiont (see our quotation from Peter Godfrey-Smith on pp. 59–60 and table 3.2). Such "vertical inheritance" of microbes, as it is called, certainly pertains in some cases. One example is the relationship between eukaryotic cells and their mitochondria: the latter, now an organelle, had its origin as a vertically inherited microbial symbiont. Another interesting case is the symbiosis between aphids and the bacterium *Buchnera aphidicola*. Let us develop the second case to illustrate vertical inheritance through a coreproduction mechanism.

Aphids, the hosts of *Buchnera aphidicola*, feed on plant sap, rich in sugars but missing certain "essential" amino acids which these insects need to build their own bodies and which they, like other animals, cannot make for themselves.[15] Their *Buchnera* symbionts are bacteria that have lost many genes necessary for independent growth but retain and express those necessary to produce and export essential amino acids to their hosts. *Buchnera* are faithfully passed from mother aphids to their progeny as embryos through special cellular structures. Thus, aphids and their *Buchnera* now *coreproduce*.

Vertical inheritance (coreproduction) can come in degrees. For example, as with insects that eat their mothers' feces, there are situations in which progeny hosts are preferentially but not exclusively exposed to the symbionts of their parents without strict coreproduction and thus exhibit something close to vertical inheritance.[16] But many systems called holobionts do not show "vertical inheritance." That is, they do not engage in collective "reproduction" as needed to define holobiont parent-offspring lineages. Instead, they exhibit what is called "horizontal inheritance," in which reproduction of a host and its microbial symbionts occur independently—and generally the microbial species are similarly uncoupled reproductively from each other as well.[17] Thus each host generation might be said to "recruit" its microbiota, species by species or cell by cell, from an environmental pool of free-living individuals (as in the left-hand half of fig. 3.1). This is the case for the most frequently studied holobionts, us humans. Those of us born "naturally" (not by caesarean section) do acquire microbes from our mothers, but many of these are replaced early in childhood. Likely only a very small fraction (if any) of the microbes hosted by a mature (reproductive) adult are direct descendants of those hosted by its parents when they conceived it.

Horizontal inheritance is also clearly the case for the squid-vibrio symbiosis, taken by some holobiont enthusiasts as a poster child for holobiosis. The Hawaiian bobtail squid (*Euprymna scolopes*) boasts an evolved "light-organ" serving, it is supposed, to make its silhouette less tempting to bottom-dwelling predators. The light shown by this organ is provided by luminescent bacteria (*Vibrio fischeri*) that are recruited by newborn baby squids from the free-living population of this species.[18] There is no reproductive coupling of host and bacterial symbiont.

So, in the human case, although many microbial parent-offspring lineages may have acquired genes aiding in their association with humans (either positively as in mutualisms

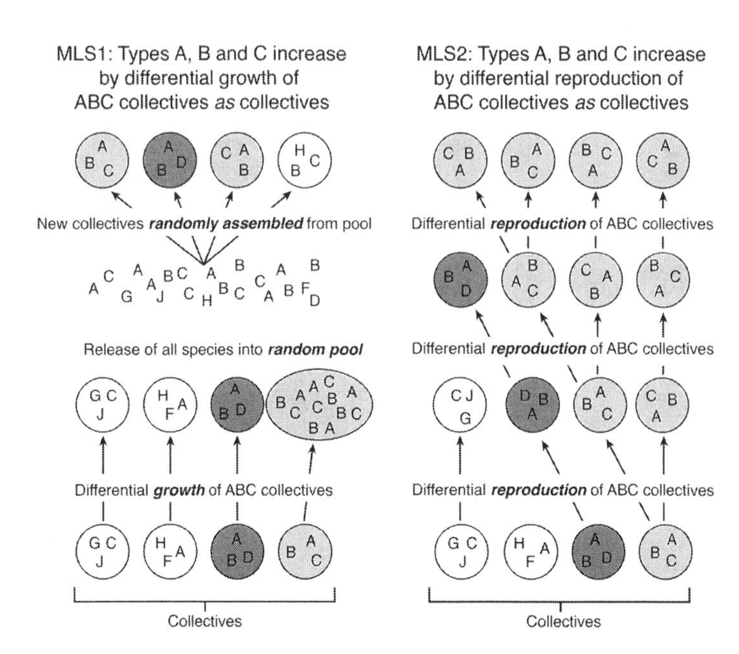

FIGURE 3.1 | MLS1 and MLS2 and community evolution. In both cases, the fraction of "ABC communities" (those with representation of the species A, B, and C) increases. In the MLS2 situation, it is because those communities reproduce (as communities) more often than do others (more often than those with other species compositions), and offspring communities tend to resemble parent communities. In the MLS1 situation, communities with representation of species A, B, and C "do better" in the sense that they produce more individuals (they get "fatter"). But at each "generation" all communities dissolve, releasing all individuals, so it's only indirectly (by increasing the numbers of A, B, and C individuals in the pool of individuals from which the next collective is recruited) that MLS1 produces more ABC communities. In either scenario there *can* be coevolution of members of species A, B, and C to increase the frequency with which they coassociate.

or negatively as for pathogens) and humans may well have in their genomes alleles selected to encourage or discourage occupation by specific microbes, all these cases can be thought of as coevolution (as defined in chapter 2). A helpful way to think about this is that each lineage (the human lineage *and* each microbial lineage, that is) retains control of its evolutionary fate and "regards" the others as parts of its environment.

Similarly in the squid-vibrio case, coevolution is clear in that squids benefit from harboring vibrios and have evolved to do this better and better, while strains of *V. fischeri* may have been selected to be hosted (protected and encouraged to grow) by local squid populations.[19] Standard ecological theory, perhaps augmented by evolutionary analyses within associated lineages, should be able to handle the analyses in either situation. By Godfrey-Smith's or Lewontin's criteria, we and squids and our respective microbes do not as holobionts show vertical inheritance or define parent-offspring lineages and so do not evolve by natural selection as holobionts, though, of course, there have been millennia of coevolution.

The rhetoric of many holobiont proponents implies much more than that, however, as if holobionts belonged as legitimate "units of selection" somewhere in Godfrey-Smith's hierarchy (genes, cells, social groups, and species, as described in our earlier quotation) and perhaps at several places in such a hierarchy simultaneously. Multilevel selection theory is taken to be relevant, but the requirement for unitary, collective reproduction it entails is seemingly ignored, at least in such recent reformulations as that of the microbial ecologist Kevin Theis and coauthors (2016):

> Discussion of evolutionary processes brings forth a second argument against the hologenome concept, namely, that holobionts and hologenomes must be the "primary" unit of selection. This strict claim leads biologists into error, as all of the literature emphasizes that multiple levels of selection can operate simultaneously. For example, selfish genetic elements

> can be selected within a genome that is in turn selected for any number of phenotypes that affect fitness—this is uncontroversial. While the holobiont is posited to be "a unit of selection in evolution," it is naturally not proposed as the only or necessarily primary unit of selection. Primariness varies with what traits are targeted by natural selection.[20]

A more clearly relevant multilevel understanding, perhaps one that Theis and his colleagues would accept, might actually be achieved through embracing the replicator/interactor framework of the late philosopher of biology David Hull, which is discussed in detail in chapter 4. But it is only the supposed intimacy of their integration, temporary though that may be, that differentiates holobionts from the biota at large—from what Charles Darwin referred to as an "entangled bank," in other words.[21] Biotic interactions are many and complex, but holobiont theory does not add a new unit to which natural selection can be said to apply.[22]

Such an objection was at the heart of the two most often cited critiques by scientists of the hologenome theory of evolution.[23] Some philosophers of biology reasoned similarly.[24] All argued that holobionts' tendencies to show correlations in species' associations is evidence of coevolution only, and often it is not even that. Evolutionary microbiologists Nancy Moran and Daniel Sloan give a humorous example,

> it has been demonstrated that showerheads are colonized by characteristic microbial communities that represent a highly selective subset of all water-borne microorganisms. This is

> expected, as showerheads provide a distinctive habitat and resources, suitable for particular sets of organisms. But showerheads and microorganisms have not coevolved.[25]

Even when all partners *are* biotic, all we need to invoke is coevolution to explain coadaptation between species.[26] This is not to say that holobiosis (mutualisms involving several microbial species in the microbiome and their hosts) cannot evolve by natural selection. Figure 3.1 (left-hand side) shows how. If a combination of members of species A, B, and C (one of which could be "the host") is beneficial in that it results in an increase in the numbers of individuals of species A, B, and C, then such interaction might be favored by selection operating individually on members of A, B, and C (coevolution). How this would work is as follows. When such interactions break up and release their contents—which is what they would do, at least metaphorically, if there is no coreproduction of the interacting individuals as a collective—then there will nevertheless be more individuals of species A, B, and C in the common pool from which the next population of interacting collectives is formed. So such productive "holobionts" will increase in numbers and relative amounts. But without collective reproduction as units, these holobionts would not be "units of selection" in Godfrey-Smith's sense: they would not replicate or reproduce as collectives and would define no parent-offspring lineages *as collectives*. There might be selection on individuals within species A, B, and C to participate in such mutualistic interactions, but this would be coevolutionary selection, each species benefiting itself. This MLS1-like process is what Roughgarden et al. cast as "holobiont selection," but it is not selection in the sense of Lewontin's Recipe.

The process is indirect but effective. Humans as holobionts are mostly like this. In such cases, the NO-REPRODUCTION problem still applies. Because such holobionts do not reproduce *as* collectives, they do not themselves evolve by natural selection, and thus they do not have properties or traits that evolved for the benefit of the collective. The evolution by natural selection occurs only at the level of the individual organisms that make them up. It could be that some microbial species function as mutualists (are good for their hosts and their hosts are good for them) but each species benefits by this. There need be no collective interest and of course many of our microbes are commensals (neither good nor bad), while some are pathogens. For regeneration (or indeed any other collective-level property), a holobiont that follows the MLS1 logic cannot be a unit of selection. Regeneration in such a case has no ultimate explanation; it serves no evolutionary function or purpose for the collective. For how could it, if the collective is not reproducing as a collective?

MLS2, on the other hand, entails differential replication of collectives *as* collectives, even though all collectives might grow to similar sizes regardless of species composition. In figure 3.1 (right-hand side), collectives containing members of species A, B, and C produce more progeny collectives as replicates. So the frequency of such collectives increases directly, and we can call such collectives "units of selection" in the sense that Godfrey-Smith, following Lewontin's Recipe, would fully accept. The aphid-*Buchnera* symbiosis is like this. In these cases, the NO-REPRODUCTION problem does not apply, but this is a rare case among holobionts and, thus, does not help us with understanding regeneration as an evolved phenomenon more generally.[27]

The Gaia hypothesis provides another good illustration of the NO-REPRODUCTION problem at work. It seems to us to involve the same faulty thinking as engaged in by more enthusiastic holobiont proponents, albeit at a very much higher or more inclusive level. In a sense, this hypothesis takes the whole world to be an interlinked "community," a giant collective mostly made up of microbes. As promoted by the renowned independent British scientist and inventor James Lovelock and the famous American biologist Lynn Margulis, the Gaia hypothesis had it that the biosphere—all life on Earth—is an "active adaptive control system able to maintain the Earth in homeostasis."[28] Lovelock and Margulis often wrote as if they meant by this that the whole biosphere *is* an organism, purposively maintaining a planet hospitable for life.[29] This notion captured the imagination of New Age enthusiasts, who deified Gaia as the Earth Goddess, but was greeted with scorn by neo-Darwinists who considered it impossible in theory.

Philosopher and historian of biology Michael Ruse writes eloquently in his 2013 book *The Gaia Hypothesis: Science on a Pagan Planet* about this notion and its reception, both public and scientific. Surely, that public intuition went, there is some "balance of Nature," at the very least some inbuilt impulse toward equilibrium or global self-preservation. In the past we might have attributed this stability to a god or gods, but now we hope to see it as the result of a natural mechanism, not supernatural intervention. Darwin had, after all, provided a mechanism, natural selection, by which the apparent purposiveness of organisms might be understood. Maybe that could be applied.

But, Darwinian theorists objected that extending this mechanism to the whole biosphere is problematic for reasons exactly analogous to those we just went through for holobionts when they do not exhibit vertical inheritance. Indeed, one of us argued, in an 1981 essay titled "Is Nature Really Motherly?" and aimed at a broad public, that failure to show *reproductive* heritable variation in fitness at the biosphere level meant that selection could only provide explanations at the level of species and below (table 3.2).[30] For explanations above the species level, "no serious student of evolution would suggest that natural selection could favor the development in one species of a behavior pattern which is beneficial to another with which it did not interbreed, if this behavior was either detrimental or of no selective value to the species itself."[31] Moreover, any global (biosphere-level) benefits contributed by any species would not be realized until long after the species in question had gone extinct, so "cheaters" that enjoyed the benefits while producing nothing themselves would be expected to arise and dominate.

A year later, Dawkins, in his book *The Extended Phenotype: The Long Reach of the Gene*, noted,

> The Universe would have to be full of dead planets whose homeostatic regulation systems had failed, with, dotted around, a handful of successful, well-regulated planets, of

> which the Earth is one. Even this improbable scenario is not sufficient to lead to the evolution of planetary adaptations of the kind Lovelock proposes. In addition we would have to postulate some kind of reproduction, whereby successful planets spawned copies of their life-forms on new planets.[32]

And still, almost forty years later, this would be the Darwinian objection. In a 2015 review of Lovelock's then most recent book, Godfrey-Smith summarized such thinking as follows:

> feedback between different living things is indeed ubiquitous, and some kinds of feedback help life to continue. But those benefits to life as a whole are by-products—they're accidental. The interactions between species are consequences of the traits and behaviours that evolutionary processes within those species give rise to, and those processes are driven by reproductive competition within each species.[33]

Compared to the holobiont debates, the Gaia controversy has had much more time to gain nuance and elaboration, and the rejection of the Gaia hypothesis by Darwinists because the biosphere as a whole cannot reproduce and so isn't an entity that can show evolution by natural selection does not mean that it has been useless as either science or politics. Both Michael Ruse and the critical biogeochemical modeler Toby Tyler (author of the 2013 book *On Gaia: A Critical Investigation of the Relationship between Life and Earth*) stress the importance of Lovelock's efforts in getting geologists to pay attention to biology. The impact of Lovelock on the development of Earth system science cannot be ignored.[34] It is not just that geology provides the environment that organisms evolve to be adapted to, but that these adaptations then change the environment—which then of course affects how organisms subsequently evolve and so forth. This process—comprising round after round of global niche construction—we might consider a form of coevolution even though the abiotic component does not enjoy evolution by natural selection.[35] Indeed, Lovelock's Gaia hypothesis was very much the product of his 1960s realization that it was the maintenance of an atmosphere in which key gases are wildly out of their equilibrium concentrations that would be NASA's first and best clue to the existence of Martian life.

But there seems no warrant, under Lewontin's Recipe or Godfrey-Smith's formulation, for seeing homeostatic tendencies at the biosphere level as adaptations evolved by natural selection at that level or favoring a sort of continual regeneration. Notions of "purpose," "adaptation," or "function" are merely metaphorical at levels above individual species, so long as reproduction is taken as a prerequisite for natural selection, as Darwinians like Godfrey-Smith or Dawkins (see our above quotation about "dead planets") do. An ecological perspective (chap. 2) in which species have coevolved through selection operating independently on each should be adequate.

In such a perspective, however, evolved regeneration of salamander tails is only superficially or metaphorically like "regeneration" of forest communities after a fire or of our gut microbiomes after a course of antibiotics. We will have failed in our mission to find a common ultimate cause in evolution by natural selection. In our next chapter, we'll try to deal with the two problems we raised above. The first, again, and posed fully in this chapter, is that microbial communities don't often reproduce as communities (NO-REPRODUCTION). The second, raised above but discussed fully in chapter 4, is that if anything is seen to be as under selection it is the pattern of interactions between species (the "song") and not those species themselves or their genes (the "singers") (SONG-NOT-SINGERS). We will argue that these are in fact the same problem and a solvable one.

4 Interactors

> Ecosystems are incredibly complex systems made up of millions of interacting parts, and their health and regenerative abilities are affected by the systems with which they interact, such as the humans around and in them. They are parts of larger biosocial systems that make up our planet. So, how does this whole system of interacting parts regenerate something that actually works?
>
> JANE MAIENSCHEIN and KATE MACCORD, *What Is Regeneration?*, 2022[1]

In this chapter we discuss how we might see microbial communities as subject to evolution by natural selection despite the two problems raised in chapter 3. It's only then that we could claim that organismal and community regeneration have similar functions or purposes in more than a metaphorical sense. We will need to modify standard Darwinian thinking as represented by Lewontin's Recipe or the writings of Godfrey-Smith, but after all not all Darwinians use exactly the same cookbook even now! We figure that what we must preserve is Darwin's original intent: to explain the adaptedness (fitness to an environment) of organisms by a reiterated error-and-trial mechanism rather than "intelligent design." And thinking about natural selection in our way, which combines multilevel selection theory with what is called the *replicator/interactor framework* of the late philosopher David Hull has many other benefits.

Recall again the first problem we must solve, NO-REPRODUCTION: communities, including microbial communities, generally do not reproduce as complete and integrated systems, so they simply cannot evolve by natural selection according to Lewontin's Recipe—even in a multilevel formulation like that of Godfrey-Smith. The second problem, SONG-NOT-SINGERS, is that often microbial community regeneration is "functional" (requiring only that functions similar to that of the initial state be performed) rather than "taxonomic" (requiring that the same species or even strains of the same species are involved as performers of these functions). We address the second problem more thoroughly now, and in the end propose a solution or solutions that, for us, can deal with both. Indeed, viewed in a certain light these two problems are *the same* problem.

For us, an early microbiomics paper provided an "Aha! moment." This 2009 paper by the computational microbiologist Peter Turnbaugh and colleagues on the "core gut microbiomes of obese and lean twins" was one of the first to set microbiomics in an ecological context.[2] As of mid-2021, it has been cited more than 6,500 times, according to Google Scholar, and was so influential in the field that quoting most of its last paragraph seems justified.

> The hypothesis that there is a core human gut microbiome, definable by a set of abundant microbial organismal lineages that we all share, may be incorrect: by adulthood, no single bacterial phylotype was detectable at an abundant frequency in the guts of all 154 sampled humans. Instead, it appears that a core gut microbiome exists at the level of shared genes, including an important component involved in various metabolic functions. This conservation suggests a high degree of redundancy in the gut microbiome and supports an ecological view of each individual as an "island" inhabited by unique collections of microbial phylotypes: as in actual islands, different species assemblages converge on shared core functions provided by distinctive components.[3]

Microbiome research has since often endorsed something like this claim—that community-level functions are more stable and predictable than is community species or strain composition. Indeed, there is a sense in which, no matter how fine-grained our definitions of function are, this just has to be true. This is because of the mechanisms by which microbiota are recruited into a community and because of the phenomenon of redundancy. What redundancy means is that there are often many species, often not even closely related species, that can perform the same ecological function and thus substitute for each other.[4]

One might argue that if members of the same species are interchangeable genetically and functionally, this is a difference that makes no difference. But traditional natural selection won't work that way. It requires reproductively defined parent-offspring lineages; in fact we cannot even expect that communities in successive generations of communities at sites with similar ecologies, or even at the same site with an apparently unchanging ecology, will comprise members of the *same species*.[5] Because of functional redundancy, it's what the organisms are doing (their function, or the "song" they sing) that matters, not their identity (who they are, or the singers), and, often, organisms of many quite distantly related species can do the same thing.

Such redundancy has been demonstrated many times over in situations in which ecologically similar sites (or the same site over time) exhibit similar patterns of ecological processes. The *same* processes are implemented by different species. From a 2018 global meta-analysis of microbiomic data, for instance, the computational biologist Stilianos Louca and colleagues conclude that,

> Depending on the choice of functions, a distinction between functional community structure and composition within functional groups can yield important insight into biogeochemistry and community assembly mechanisms. Indeed, metabolic pathways involved in energy transduction can be strongly coupled to certain environmental factors and elemental cycles, and can appear decoupled from particular taxonomic assemblages. Similar observations are known from macrobial ecology, which has had a long history of describing community structure in terms of guilds, life forms and strategies, all of which may be considered analogous to metabolic functional groups in microbes.[6]

Redundancy can, of course, simply reflect phylogenetic relatedness, as it does in the ecology of "macrobial" systems. An ecosystem in which there are plants and large herbivores could attract wolves, coyotes, or mountain lions, for instance, but these are all members of Carnivora, after all, related at some level and sharing a common meat-eating ancestor. At a much finer scale, sister microbial species (those sharing a very *recent* common ancestor) are more likely to boast similar metabolic capabilities and perform similar ecological roles than are two species from wildly different groups. Often indeed, it *will* be different species of the same genus, or different strains of the same species, that replace each other, performing the same community function at similar sites, or the same site over time.[7]

But the microbial world has an additional trick up its sleeve, so to speak, which can radically disconnect phylogenetic relatedness and redundancy. Often, functional redundancy reflects lateral gene transfer between species that are distinct from each other, even effectively unrelated on any Tree of Life. Indeed, one of the great surprises of comparative genomics as it has developed over the last few decades is the extent to which microbial genomic composition is the product of lateral gene transfer (fig. 4.1). And thinking about genes as transferable entities rather than just eternal parts of the chromosomes of some particular organism or species encourages the expansion of evolutionary theory we suggest later in this chapter.

For this reason, we turn now to an expanded discussion of the importance of lateral gene transfer within microbial commu-

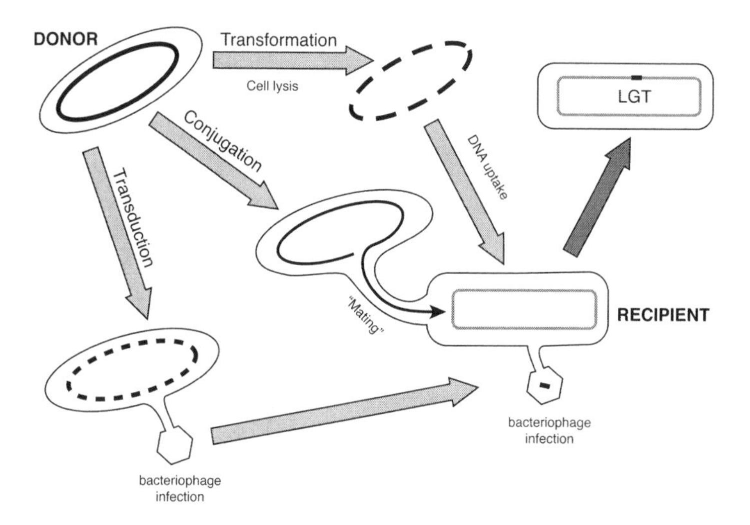

FIGURE 4.1 | Processes contributing to lateral gene transfer. Transformation entails the lysis (physical dissolution) of a donor and the uptake (and genomic incorporation) by the recipient of fragments of its genome. Conjugation ("mating") involves the elaboration by the donor of a conjugative apparatus and results in transfer to the recipient of all or a portion of the donor's genome and/or independently replicating elements that the donor cell might bear (antibiotic resistance elements or integrative conjugative element-like elements as in the text). Transduction is the (usually accidental and usually rare) incorporation by an infecting virus (a "bacteriophage") of a fragment of donor DNA (often at the expense of the viral genome) and its injection into a recipient cell. Various mechanisms or simple recombination can integrate the introduced donor DNA into the chromosome of the recipient. These three are not the only processes now known to effect lateral gene transfer.

nities. We've referred to this process often in the preceding pages and it's now time to say a bit more about what it is and how it occurs. Lateral gene transfer, as we'd define it, is any genetic mechanism by which a gene or genes from one species becomes part of the heritable genetic complement of another species which previously did not have that gene or genes, or at least not the same versions of that gene or genes. Lateral gene transfer is common in microbes, much more common than we thought a few decades ago. We do not include homologous recombination, which is the exchange of pieces of genes of very similar sequence among members of the same species. This last is roughly analogous to what happens with our own human progeny who inherit some genes or parts of genes from their mother and some from their father.

Lateral gene transfer is much more radical in its effects than that. Although bacterial "species" are admittedly hard to define (as we've discussed throughout this book), lateral gene transfer can occur between groups that are so far apart evolutionarily that no sensible microbiologist would ever consider them the same species. There are even instances in which lateral gene transfer crosses boundaries between "Domains," from bacteria into animals, for instance, as discussed below.

Lateral gene transfer is often inferred to have occurred from phylogenetic analysis. This involves the comparison of sequences of the same genes in different species, performed in order to establish the evolutionary relationships of those species and place them in the Tree of Life. In such analyses, if we find radically different family trees for different genes in the same species' genome and can't dismiss it as some sort of experimental or computational error, then it *must* be that one or the other gene came from some other species. Lateral gene transfer can also be revealed by wildly discordant gene properties, such as nucleotide composition (relative frequencies of the bases A, T, G, and C). Nucleotide compositions tend to become uniform across a genome with time, and recently transferred genes may show different compositional characteristics just because their donor had a different nucleotide composition. Simple presence/absence data can also be used.[8] If a gene is only present in a single strain of a species, then it's simpler to assume that it was added to that strain rather than lost from all the others.

A few years ago, for instance, David Ussery, a computer scientist interested in microbes, and his coworkers described results from comparing the sequences of more than 2,000 strain isolates of the common human gut-dwelling microbe *Escherichia coli*.[9] On average, each strain's genome carried about 5,000 genes, but the "core" (shared by all or most of the strains) was only about 3,100 genes, while the "accessory" set, found in just some (as few as one) genomes was 89,000 genes (and counting). It's likely that there is relatively rapid turnover of the latter set, involving roughly balanced gain by lateral gene transfer and loss by mechanisms that delete stretches of DNA. We have to assume a rough equilibrium in bacterial genome size, after all. Attributing strain differences to loss only would imply that the ancestor of all the *E. coli* strains had at least 89,000 genes, 10–20 times as many as any known bacterium does now!

Sometimes the rapidly-turning-over genes will be parts of obviously self-replicating "selfish" DNAs of viral or other mobile element origin (like the integrative and conjugative element described in the next paragraph). Such elements make their living by replicating independently from chromosomes and being transferred together between species, often at the expense of organisms within those species. But sometimes such elements will also carry genes that clearly can benefit the recipient organism and determine adaptive responses to local environments, such as antibiotic resistance. In fact, microbial "speciation" events can be initiated by acquisition through lateral gene transfer of a relevant gene, and such transferred genes will sometimes clearly be derived from other microbes already living in such environments.

As an example of this, the microbiologist Jan-Hendrik Hehemann and colleagues found, a decade ago, that some Japanese people carry in their guts a strain of the symbiotic bacterium *Bacteroides plebeius* boasting genes for carbohydrate metabolism that were transferred from an ocean-dwelling bacterium that metabolizes similar compounds on marine red algae like *Porphyra*.[10] Such algae include species used in the making of *nori*, frequently eaten in Japan. The advantages to *B. plebeius* and its human hosts seem obvious. The mechanism of transfer of at least some such genes in this case is what's called an integrative conjugative element.[11] Such elements carry the genes for inserting themselves as a stretch of DNA into a recipient bacterium's chromosome and, also, when triggered, for excising themselves again, producing the machinery and structures necessary for conjugation (fig. 4.1) and transferring to a cell of another species.

The "purpose" of carrying extra genes useful to hosts, like those involved in algal carbohydrate metabolism, can easily be rationalized in terms of the differential replication of recipient gut bacteria and the integrative conjugative element-like entities they contain. Think of these beneficial genes as offerings to the host: the more environments that support the growth of *Bacteroides plebeius*, the more copies of this particular integrative conjugative element there could be. Moreover, to the extent that Japanese people who can derive energy from eating nori are advantaged over those who can't, we might want to see this as an adaptation at the level of the holobiont—that is of a human together with his or her microbiome (as discussed in chapter 3). So genes for algal carbohydrate metabolism might be said to confer selective advantage at many levels—the level of the integrative conjugative element, that of the integrative conjugative element-bearing *B. plebeius,* and that of its human host. Moreover, to the extent that Japanese societies have benefited by being able to derive more energy from eating seaweed, we could see nori-digestion as an adaptation at that cultural level, too, although perhaps that's a stretch.

Bacteria are *prokaryotes* (which means cells that have no nuclei) and mostly single cells. Most of the animals and plants that we can actually see are multicellular *eukaryotes,* organisms made up of eukaryotic cells (having nuclei and many other complex cellular structures). Lateral gene transfer into animal and plant genomes (from bacteria or other eukaryotes in their niches) is presumably less frequent than it is among bacteria, if for no other reason than that many animals, like us, have sequestered "germ lines." The only way that a gene laterally transferred into an adult human can make it into that human's progeny and thus into their progeny and be evolutionarily significant would be to be transferred directly into a cell destined to become an egg or sperm. Transfer into some intestinal cell, say one lining the gut where microbes are most often encountered, won't do it. So claims that we humans have a lot of genes acquired from bacteria since we diverged from other primates are probably untrue.

But there are some robustly proven instances of lateral gene transfer into eukaryotes, especially into single-celled protists (protozoa), which don't have separated germ lines. And these cases are often easily interpreted as the acquisition by the eukaryote of genes it needs to get along where it is. For instance, computational biologist Laura Eme and colleagues show that species of *Blastocystis,* protists common in human guts, have acquired 2.5 percent of their genes by lateral gene transfer from bacteria and other eukaryotes.[12] Among these are genes for carbohydrate scavenging and metabolism, anaerobic amino acid and nitrogen metabolism, oxygen-stress resistance and pH homeostasis, and, potentially, escaping host defenses, all useful in the gut. Lateral gene transfer has enabled *Blastocystis* to invade our guts and is probably more prevalent still among gut bacteria. Thus, lateral gene transfer cannot help but be useful, directly or indirectly, in restoring functions to a gut microbial community devastated by, say, antibiotic use.

There are also many instances in which the frequency of lateral gene transfer seems to increase in situations in which transfer might usefully rescue strains or species lacking genetic determinants necessary for survival in otherwise unsuitable or hostile niches. Indeed, lateral gene transfer was first detected in the 1960s by infectious disease microbiologists observing the spread of antibiotic resistance to previously sensitive bacteria in Japanese hospitals and is now responsible for the worrisome increase in "superbugs" associated with the profligate use of antibiotics.[13] Often, sublethal antibiotic exposure, by inducing a genetic stress response, "mobilizes" integrative conjugative elements or other genetic elements carrying multiple genes for antibiotic resistance.[14] That's one reason why we are told to use up our antibiotic prescriptions.

So, this is all to say: lateral gene transfer can be a cause of the redundancy that underwrites the regeneration of function in a community. And there's still another reason for focusing on it. We often consider genes as "belonging" to the organism whose genome they occupy. But when that occupancy is clearly transient, or easily shared with other species as redundancy through what lateral gene transfer makes possible, and when benefits accrue not only to organisms but to more inclusive entities like species or even larger and more heterogenous groups to which the genes belong (metabolic "guilds" defined by the possession of similar capabilities), such an organism-centric view seems arbitrary.

It's worth going a little bit further with this way of thinking. We'd expect frequently transferred genes to bear "adaptations" that can facilitate their transfer between hosts. The integrative conjugative elements mentioned above clearly do that. Such adaptations might include the ability of the gene's protein product to function independently of other cellular proteins and thus to function in a variety of cellular backgrounds (that is, different hosts). The mechanisms of lateral gene transfer (fig. 4.1) generally only result in the transfer of one donor DNA fragment to any recipient, generally bearing at most a dozen genes or so, so this imposes selective constraints.

Indeed it is the case that genes whose products *must* function together (catalyzing successive steps in a biosynthetic pathway, for instance) are often tightly linked in what are called *operons*—tight clusters of functionally related and coordinately regulated genes—whose existence might be seen as the result of *selection for transferability*.[15] This ingenious idea of the microbiologists Jeffrey Lawrence and John Roth is called the "selfish operon hypothesis." They reasoned that a single gene determining the performance of a single step from a multistep biosynthetic pathway (let's say the five steps involved in formation of the amino acid tryptophan) is useless by itself. Thus there will be selection for the five genes to be close together on chromosomes (as the tryptophan biosynthetic genes most often are). That way, a single lateral gene transfer is more likely to carry them all. Sometimes this is true: so natural selection for transferability itself does occur.

There should also be truly cosmopolitan solitary genes not committed to serving the phenotypic fitnesses of any particular organism or even species. Antibiotic resistance genes in "superbugs" would be a prime example. Notably, this would be an alternative type of selection, not necessarily favoring genes that enhance the reproductive potential of organisms within a species' population but instead favoring genes that, regardless of such effects, are better at hopping from one species to another, thus increasing their own representation in the biosphere. They might even sacrifice, so to speak, guaranteed membership in one limited population (that of a single host species) in favor of potential membership in a much larger (multispecies) population.

Thus lateral gene transfer offers a new way of thinking about the fact that for microbial communities it may often be that collective "function" that is more conserved and more fitted to the ecological niche than is taxonomy or classification (that is, community composition assessed in terms of species or strains). Lateral gene transfer may help encourage the selection of truly cosmopolitan genes, ultimately favoring functional redundancy within a community. It may ultimately be genes and the functions they determine, rather than the species that house them, that really matter when it comes to how natural selection affects microbial community structure and change.

Functional redundancy makes for resilience at the community level and, if functional restoration is what we mean by community regeneration, redundancy is also vital to our effort here. The greater the number and diversity of species with the same genes and performing the same functions the better. This addresses the second problem we've been considering (SONG-NOT-SINGERS) as it suggests that there are strong reasons in favor of treating community regeneration as fundamentally about the restoration of functional capacities. From the point of view of the community, so to speak, what matters is that different species are able to perform the same function, and lateral gene transfer may encourage this. This fits well with a similar conclusion arrived at by some scientists studying organismal regeneration, such as developmental biologist Richard Goss, quoted in chapter 1.[16] It's how the regenerated limb or organ functions, not its anatomical structure or the genes that produce it that matters.

But now at last we must deal with the first and possibly thornier problem we've been putting off in our disquisition about functional redundancy and the processes that underwrite it: the NO-REPRODUCTION problem. This problem, again, is that communities, including microbial ones, do not "reproduce" *as complete and integrated systems* and so cannot evolve by natural selection—even in a multilevel formulation like that of Godfrey-Smith. Even if what really matters is function, it's the organisms bearing those functions that are able to reproduce: genes alone can't do it and communities don't. Functions might be re-produced, but they don't reproduce.

So we have to rework our understanding of evolution by natural selection if we want to put holobionts, Gaia, and salamander limbs all on the same footing. But doesn't solving the second problem make solving the first even harder? Not only is there no collective reproduction of all the organisms within a community, but even species identity is not necessarily conserved!

Fortunately, there is a way to address these problems. During the 1980s, the philosopher David Hull sought to generalize Richard Dawkins's gene-centric view and our solution is based on Hull's thinking. But before we present it, we need to say more about the gene-centrism Dawkins emphasized. That view, as persuasively argued in his enormously popular 1976 book, *The Selfish Gene,* was that genes, originating perhaps as self-replicating RNA molecules some 4 billion years ago,

> now swarm in huge colonies, safe inside gigantic lumbering robots, sealed off from the outside world, communicating with it by tortuous indirect routes, manipulating it by remote control. They are in you and in me; they created us, body

> and mind; and their preservation is the ultimate rationale for our existence. They have come a long way, those replicators. Now they go by the name of genes, and we are their survival machines.[17]

The "tortuously indirect" remote controls were what molecular biology uncovered in the 1960s and 1970s, leading to the phenotypes on the basis of which organisms (the "lumbering robots" Dawkins also described as "vehicles") are selected. That discipline showed us, in what is now amazing detail, how DNA → RNA → protein (the so-called Central Dogma). Genes are the guarantors of heredity, variation and (through their expression in phenotype) fitness, and it's they that form long-lasting lineages, replication after replication. So they ensure their own perpetuation by elaborating organisms ("lumbering robots" or "vehicles") that interact with their environments (including others of their own kind) in such ways as to have more progeny.

Of course, in sexual species progeny receive only half their genes from each parent, only (on average) a quarter from each grandparent, and so forth to previous generations. So, what we humans call *reproduction* is not actually *replication*, even though the two words are often taken as synonymous. We might "reproduce our own kind," but we don't make clones of identical progeny because each of us contributes only half the genes that wind up in an offspring, only on average one quarter that wind up in a grandchild, and so forth. The relevant criterion of Lewontin's Recipe nevertheless does apply. That is, offspring of a particular pair of parents do *tend to resemble* each of those particular parents more than they do the parents of other offspring or the offspring of other parents.

Still, because of the two-parent ancestry that goes with sex and the fact that our chromosomes freely recombine and scramble the results of previous matings, any advantage in differential reproduction that a gene might confer travels with the gene in the long run. It does not in the long run travel with any particular organismal lineage as defined by iterated "parent-offspring" relations. Counterintuitively, perhaps, each of us has biological ancestors some generations back (parents of parents of parents . . .) from whom we inherited no DNA, and each of us will have great-great-great . . . grandchildren (the number of "greats" is hard to predict) to whom we will leave no DNA.

This is not a problem for asexual organisms or indeed any entities that have only one parent: genes and the phenotypes they determine *do* travel together. For such entities, reproduction is effectively a sort of replication. Clearly this is true for genes, cells, and organisms in mostly asexual species like bacteria, and arguably it's true for species themselves, if speciation most often involves a single ancestral species splitting into two.

In writing *The Selfish Gene*, Dawkins focused down on phenotype-determining genes in sexual organisms like ourselves, though. He, like most architects of traditional Darwinian thinking, is a zoologist, not a microbiologist, and was almost forced into the gene centrism his book made so famous. The philosopher David Hull's helpful generalization, on which our solution will be based, was to recognize (or expand on) Dawkins's understanding that the "lumbering robots" ("vehicles"), which are *organisms varying in phenotype* and the *genes that make them*, are, respectively, *interactors* and *replicators*. In 1980 he formalized this two-part "unit of selection" as follows:

> *replicator*: an entity that passes on its structure directly in replication
>
> *interactor*: an entity that directly interacts as a cohesive whole with its environment in such a way that replication is differential
>
> . . .
>
> *selection*: a process in which the differential extinction and proliferation of interactors cause the differential perpetuation of the replicators that produced them[18]

One might wonder how Hull's two-part "unit of selection" relates to our discussion of units of selection in the last chapter. Hull's replicator/interactor framework separates Godfrey-Smith's "units of selection" into two mutually dependent parts: that which is selected as a matter of bookkeeping (the replicator) and that which is *the target of* selection on which the environment impinges (the interactor). Asexual entities (genes, cells, bacteria and other asexual organisms, and even species, either sexual or asexual) can all be seen as replicators. These are entities that pass on their structures more or less directly in replication.

Interestingly though, each of these can also function as an interactor (fig. 4.2), insofar as interaction with its environment (biotic and nonbiotic) determines its fate. Genes, for instance, must interact with the cellular DNA replication machinery and

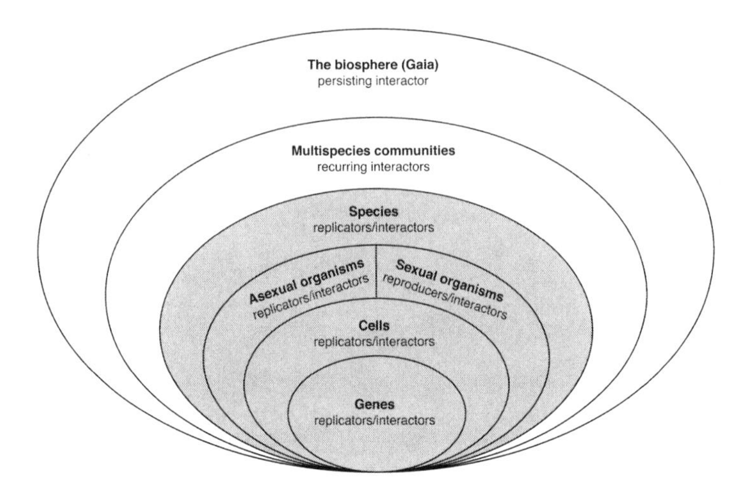

FIGURE 4.2 | Another hierarchical representation of the organization of life, this one meant to accommodate the replicator/interactor framework of David Hull. Genes, cells, and asexual organisms are all replicators in the sense that they make more or less accurate copies of themselves. They are also interactors in the sense that their survival and replication require successful interactions with biotic and abiotic features of their environments (cells in the case of genes, multicellular organisms for cells, communities and hosts for asexual organisms like bacteria). Speciation of species (asexual or sexual) is a form of replication insofar as new species are derived from existing and largely similar ones, and species interact with their environments. These interactions, in part, determine how often species speciate or go extinct. Multispecies communities (including microbial ones) often do not replicate as communities, and yet to the extent that similar communities are formed in similar environments, they "recur" or are re-produced. They also interact with their biotic and abiotic environments, such interactions being the subject matter of the new science of microbiomics. Earth's biosphere (the biotic part of Gaia) has persisted for four billion years and is arguably subject to natural selection, even if not replicating. The interaction between the biotic and abiotic aspects of Gaia seems crucial to their joint evolution. Further discussion of the meanings of "replicator," "reproducer," "interactor," and "persisting" is found in the text.

bacteria with their environments and hosts. In sexual species, Dawkins's "selfish genes" do this too (interact with cellular replication machinery), but, more importantly for evolutionary theory, they also determine organismal phenotypes, thus producing interactors whose "differential extinction and proliferation" via organismal reproduction cause such genes' differential perpetuation through replication.

In general, Hull thought, as do we, that although replicators which are also interactors could be found at lower levels of the biological hierarchy, most communities and higher-level collectives could be considered *only* interactors (fig. 4.2). This includes humans and their associated microbiota. Some holobiont proponents who understood how those holobionts showing horizontal inheritance violate Lewontin's Recipe do embrace something like Hull's replicator/interactor framework but claim that "holobionts [of both sorts] can be considered levels of selection in evolution because they are well-defined interactors, replicators/reproducers, and manifestors of adaptation."[19] But this seems to miss the essential point that many holobionts (we humans, for instance, perhaps the most talked-about "holobionts") are *not* replicators but *only* interactors.

Similarly, advocates and practitioners of the new and promising research agenda called "community and ecosystem genetics" (a field that investigates the genes that determine community, rather than organism, traits) seem to embrace multilevel selection theory but either do not require that higher levels such as communities replicate or reproduce as communities or do accept as relevant evidence of "community heredity" experiments in which replication is artificially enforced by the experimenter.[20] Neither seems to us as likely to achieve the unified explanation of organismal and community regeneration as is a full-hearted adoption of Hull's replicator/interactor framework, which we will now elaborate and combine with the multilevel selection approach explained in chapter 3 (see table 3.2).

Remember that Dawkins, in the *Selfish Gene,* focused on the role of genes in determining organismal phenotype, seeing these as the only "units of selection" (in the sense of Godfrey-Smith), at least for sexual species. This was an enormously useful and powerful rhetorical maneuver and, in line with a tough-minded insistence on reductionist approaches, embraced by many evolutionary biologists then and still. Biologist Tom Whitham and his colleagues, advocates of the community genetics framework referred to above, for example, cite the very influential evolutionary biologist George Williams as an originator of such thinking, noting that "Williams's thesis was greeted with certainty by most behavioral, ecological, and evolutionary biologists, a consensus that largely persists today."[21]

But, as Whitham's own discipline of community genetics has amply demonstrated, the phenotypic expression of many genes extends many hierarchical levels above that of the organism. As an example from earlier in this chapter, particular genes in particular bacterial species in the guts of some Japanese people have a role in their human host's ability to eat nori, and this can legitimately be said to be part of their extended phenotype: the holobiont, comprising a human and the human's microbes, functioning as an interactor that causes the differential perpetuation of the replicators (those genes). At a higher level still, the large community of human holobionts now engaged in the culture of nori-eating (enthusiastically embraced by many North Americans) could also be described as an interactor and, if the altered strain of *B. plebeius* were to make it across the Pacific, could contribute to the perpetuation of the relevant genes. If one also accepts that cultural practices themselves sometimes evolve by natural selection (as did Dawkins), then the practice of nori-eating itself might also be said to be selected for.

Whether or not higher-level interactors reproduce (like we do) or are only re-produced (like a microbial community or any other multispecies ecosystem or even the biosphere), so to speak, doesn't matter. Communities do not need to reproduce to perpetuate replicators—they can simply recur (be re-produced) more frequently or show greater stability in their interactions with their biotic and abiotic environments in order to perform this function.

Assume for instance that healthier humans or other so-called holobionts last (persist) longer than unhealthy ones and also that throughout their longer lifetimes these healthy holobionts excrete more microbes which thus (together with their genes) have some greater chance of finding themselves in future humans or other holobionts. If both are the case (as they almost certainly are for us), then microbial species and microbial genes that contribute to holobiont health will themselves be perpetuated and increase in numbers. If, on the other hand, our microbes were acquired at birth from an environmental pool and never left our bodies again (being buried with us), there'd be no selection on them to make us healthier. There'd be selection on us to choose to acquire at birth only microbes that made us healthy but none (or even negative selection pressure) on them to be so chosen.

Such a system can, of course, be exploited by the unscrupulous and often lethal *Vibrio cholera*, for instance, that perpetuates itself by causing its hosts to have vibrio-spreading diarrhea. But care must be taken by any such pathogen not to kill its host and thus end its own spread. Thus do diseases like the infamous Australian rabbit-infecting myxomatosis virus (and SAR-CoV-2, one hopes) often (though not always) "attenuate" themselves. Host-beneficial (or in any case less detrimental) microbes are expected to do better in the long run.

So, with Hull's framework, we don't need communities to reproduce together as communities in order to see how they could evolve as an end result of natural selection. It is enough to imagine that functionally similar communities will reassemble after disruption (certainly "disruption" includes host death) and that such reassembly benefits ("differentially perpetuates") the genes that promote such reassembly.

We could then see, and this is our principal point, that organismal regeneration and community regeneration have the same ultimate cause—selection for restoration of structure or function. For organisms, "purposive" (as opposed to "accidental") regeneration results from organisms that have such an ability having more progeny than those that don't. In the long run, using Dawkins's gene-centric view, it's the genes that produce regenerated organismal structures that are perpetuated and come under selection. For microbial communities in a replicator/interactor framework, it's the replicators (provisionally just genes, but see below) that produce resilient communities, or communities that readily recur whenever conditions are right, that are perpetuated. Hull's replicator/interactor framework, supported by more recent work on lateral gene transfer, can help us think through the ultimate explanation of both organismal and community regeneration and in common terms. The "why" of organismal and community regeneration are, in Hull's framework, the same. Either system regenerates in order to ensure the perpetuation of the replicators that determine its interaction with the environment. That is regeneration's purpose or function, and it's more than an accident that it occurs in both contexts.

There are several additional things to say about the modifications to traditional Darwinian theory entailed by combining multilevel selection theory and Hull's replicator/interactor way of thinking. One is that the analogy to sexual reproduction that in part drove Dawkins to gene-centrism goes deep. If it is genes that produce community interactions (phenotypes), these genes need not always be in the same genome or even genomes of the same species. Analogously, in our two-parent inheritance system, recombination between and within chromosomes means that the genes favoring limb regeneration need not travel with genes favoring other aspects of organismal phenotype. Put another way, if multiple genes are required in either organismal or community regeneration, these need not always originate in the same genomes or (for communities) the same species. Successful regenerators at either level are interactors whose success (the avoidance of extinction or frequent environmental recurrence) causes the "differential perpetuation" of the genes that they recruit, either through genetic recombination (for organisms) or through ecological assembly (for communities).

We do not need reproduction for natural selection to occur as long as we have *re-production* (with a hyphen), by which we mean *producing again*, reassembly through ecological recruitment as above, or resilience of a particular community, in other words, *differential persistence*.[22] Interactors can become better adapted to their environments without needing to be selected on the basis of their own differential reproduction.

We also see this way of thinking as capitalizing on the distinction between MLS1 and MLS2 as was shown in figure 3.1. Multilevel selection theory, as exemplified in the quotation from Godfrey-Smith in chapter 3, demands that there be reproduction at any level said to be "under selection," any level harboring "units of selection." In other words, that something like MLS2 applies. The replicator/interactor framework, on the other hand, accepts MLS1 as a legitimate process by which genes might, through allowing the differential growth (or persistence) of the interactors they form, serve their own ultimate interests, so to speak. Collectives in the left-hand side of figure 3.1 play the role of interactor, and the genes favoring community growth are the replicators. Godfrey-Smith, whom we've used as a foil for traditional views here, would actually agree with this part, we think. In his 2014 book, *Philosophy of Biology*, he notes that "the other part of [David Hull's] framework, the idea of an interactor as an evolved object, might be useful in dealing with symbioses and the like. There are objects that recur in evolution without reproducing as units. Their parts reproduce, and the parts come together to make more of these recurring objects."[23]

We also note, and this is a significant point given the commonly assumed reductionism described above, that although one can always reduce selective advantage as something ultimately accruing to genes, it may not be necessary to do that. Table 3.2 indicates that all four of the lower levels in our hierarchy comprise units that are both replicators (or in the case of sexual species, reproducers) and interactors, while the two higher levels are made up of entities that are only interactors (in the case of Gaia, arguably in a population of one). A broad construal of Hull's replicator/interactor framework might have any interactor at any level "cause the differential perpetuation" of replicators/reproducers at any lower level. Thus, the differential recurrence of multispecies communities could cause the differential perpetuation of species capable of participating in such communities—wolves, coyotes, or mountain lions, in our earlier example, or microbial species sharing a common metabolic capacity, possibly as a consequence of lateral gene transfer of the relevant genes. So in our discussion of MLS1 above (as in fig. 3.1), we might instead cast the individual organisms containing community-beneficial genes, or the species containing such organisms, as the relevant replicators.

For organismal regeneration, a similar relaxation in what is the replicator might be entertained. The genes whose differential expression causes regeneration of organismal structures need not be those involved in the initial development of such structures, or necessarily conserved between species, for such structures to be thought "the same." It's the structure that is replicated. Earlier, we cited examples from amphibians and lampreys. Selection for regeneration in organisms and for recurrence of similar microbial communities are, in replicator/interactor formulations, the same process, ultimately speaking.

So, we think, our difficult goal can be accomplished, at least in theory (see the epilogue). There is an enormous amount of empirical work to be done. Microbial communities, especially that in our gut, are amazingly variable from person to person, and the literature on community "regeneration" after a disturbance such as antibiotic treatment also shows enormous variation. Meanwhile, a 2021 review of higher-level community resilience concludes that, to date, "our ability to predict the direction and magnitude of microbial and ecosystem responses to global change has not really improved."[24]

Our goal here was a philosophical one: that we might address organismal and microbial community regeneration in the same evolutionary terms. But there are practical implications for ongoing research. Adoption of the replicator/interactor framework and recognition that both interactors and replicators can be found at levels higher than those of organisms and the genes that make them means that we must search for those replicators and flesh out those interactions. A renewed emphasis on establishing the community-level (rather than organism-level) function of specific genes is one benefit of such refocusing. There may well be genes that benefit community functions while being (slightly) detrimental to the organisms that temporarily bear them. And it may be that lateral gene transfer has itself been selected for as a mechanism for maintaining community function rather than through the selfishness of the agents of transfer. We need not only gene-based evidence for this but realistic computational modeling to show that it is feasible.

5 Engineering

It is sobering that we still know so little about ecological communities that we cannot reassemble them with anything approaching real success. [. . .] The current limits of knowledge, coupled with the daunting complexity of ecological systems, make it all the more imperative to study and preserve the natural communities that support the basic ecosystem functions on which we depend. There is much left to learn [. . .] and that potential knowledge will be lost forever if we allow natural communities to disappear before we learn their secrets.
PETER MORIN, *Community Ecology*, 2011[1]

Humans, industrious creatures as we are, seldom settle for merely understanding nature. Our aim is often to put our knowledge in the service of engineering: to mold nature into systems that work for us. The complexity of nature demands that this involve a dialogue, to use a metaphor, between the natural and the artificial: we study natural systems to learn their secrets, we attempt to apply these in the engineering of new systems, then we turn back to nature for further secrets, and so forth. How we engineer microbial communities will be the focus of this chapter, but as a way of connecting this to our previous discussions of ecological and evolutionary theory, we begin with a review of what we've presented so far.

Ecologists use the term "regeneration" broadly, and we have followed suit. Regeneration is the reemergence of a taxonomically or functionally similar community of organisms following a disturbance. Explaining why regeneration occurs has involved a discussion of proximate explanations—which appeal to the sequence of causes that bring about regeneration—and ultimate explanations—which appeal to the function or purpose regeneration performs.

Appealing to natural selection to offer ultimate explanations of function and purpose is commonplace in and central to evolutionary biology. Recall that the presence of a turtle's shell can be explained by the function the shell has performed, namely, protecting turtles from predators. Similarly, the regeneration of the salamander limb can be explained by the function it has performed: it is allowing salamanders to regain mobility following limb damage so that they can continue to have baby salamanders (perpetuating the genes required for limb regeneration). The trick for microbial communities was explaining how (and under what conditions) communities can evolve by natural selection. The problem was that most microbial communities don't reproduce in any useful sense of that term—they don't have "baby communities"—and so do not conform to standard understandings of evolution by natural selection. Broadening those understandings allowed us to close a theoretical gap between organismal and community regeneration. The solution we proposed is applicable as well to the much more familiar communities made up of visible animals and plants, the Georgia piedmont, for instance—another community that does not reproduce as a whole.

And our solution raises an interesting question about conservation biology: is it really functional interactions rather than particular species we are trying to preserve? Few would care if some bacterial species like the *Bacteroides plebeius* mentioned in chapter 4 were to go extinct as long as another was ready to take its place, and because of redundancy, there might well be. But many care about the fate of the polar bear (*Ursus maritimus*). Is it only nonredundancy (there aren't all that many cold-adapted large four-legged predators to choose from) that makes these cases different, or are there other values at stake?

With discussions of proximate (ecological) and ultimate (evolutionary) explanations concluded, it may seem then that our goal is met and our job done. But as noted in chapter 1, making that assumption would be to overlook a different and important way in which ultimate explanations and invocations of "function" and "purpose" can be appropriate when explaining why systems of all kinds have the structure they have or behave in the way they do. Put living systems aside for a moment and imagine a bicycle. Why does a bicycle have a seat? An appropriate ultimate explanation, appealing to the function of the seat, is that a bicycle possesses a seat because it allows the rider to sit comfortably. That is a perfectly acceptable answer to a why-question that draws on function. But note that this isn't (necessarily) an evolutionary story: bicycles don't reproduce, and it might be difficult to see them as evolving by natural selection.[2] This is a different kind of ultimate explanation, one we will call an engineering ultimate explanation.

When systems, both living and nonliving, are designed and engineered by humans for a particular purpose or function, an ultimate explanation is also warranted, one aligned with the human designer's intent. It's this intent that gives purpose and defines function.

Like the bicycle, living systems, such as ecological communities, can also have *engineering-type* ultimate explanations when humans intentionally design them to perform special functions. They are structured the way they are and they behave the way they do (one might even go so far as to say they exist at all) because of the functions they perform for us. Community regeneration should sometimes be explained in this way. As we've mentioned already, one reason why so-called monocultural fields, those containing only corn, continue to reemerge each year throughout the midwestern United States following harvesting (that is, after an ecological disturbance event caused by us) is because these communities serve functions for the industrial food system. Why else would we see the perpetual regeneration of corn fields? Their regeneration can, in other words, be explained by the function it serves for humans. They have been engineered to provide a reliable and abundant food supply for us or our animals.

And beyond explanation, engineering also importantly represents a different kind of scientific endeavor from the ecology and evolutionary biology we've been describing in earlier chapters. Our discussion of proximate and ultimate explanation was aimed at understanding what we know about *how* community regeneration works and *why* it occurs. But the engineering perspective is slightly different: its aim is what we can *do* with this knowledge. How can we put this knowledge in the service of making pathological communities healthy again? So, in much of what follows in this chapter, we show how the lessons learned in the previous chapters have been used to further engineering goals.

The engineering of microbial communities leading to their regeneration is central to sustainability science applications and in medicine: humans attempt to design and engineer specific microbial communities that are helpful for our lives, and we attempt to regenerate the functions of these helpful communities following disturbances.[3] In this chapter we discuss the engineering of microbial communities so that they help humans. Following standard terminology, and explained further below, we call helpful communities "eubiotic," distinguishing them from harmful "dysbiotic" communities. We focus specifically on the human gut microbiome (the community of microorganisms in the human gut) because it is central in the popular media and medical sciences and because there is a lot of research on the strategies that are effective for regenerating eubiotic communities in the gut (fig. 5.1).

Let us begin with the goal of engineering. This is often the regeneration of a eubiotic or healthy community, but what does that mean? One popular example of purported eubiosis might be how maternal microbiomes are acquired by infants born "naturally" (vaginal delivery) versus by medical intervention (C-section). Infants born naturally show a different composition of gut microbes than infants born by C-section. And because there is evidence that the latter practice, as well

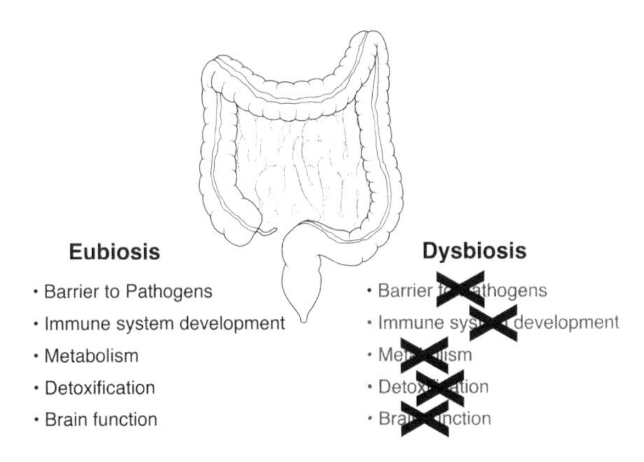

FIGURE 5.1 | The functions of human gut microbiota in eubiosis and dysbiosis.

as formula- rather than breast-feeding and early antibiotic use, puts infants at risk for developing celiac disease, asthma, type 1 diabetes, and obesity, natural birth is thought to give rise to a eubiotic community, whereas C-section a dysbiotic one.[4] This example also has an engineering component aimed at restoration of eubiosis from a dysbiotic state: one way to help infants born by C-section is deliberate exposure to the vaginal microbiota of their mothers.[5]

The terms "eubiosis" and "dysbiosis" gained popularity in the literature about holobionts, as discussed in chapter 3. This literature has generated a broad framework within which the relationship between animals or plants and the microbes on and in them is most commonly conceived. But within this framework, the terms eubiosis and dysbiosis are not always used in the same way and sometimes carry with them unstated and problematic ecological and evolutionary assumptions. For example, it is sometimes assumed that microbial communities have coevolved with their hosts to produce multispecies mutualisms, that is, symbioses that are good for all involved without evidence confirming that this is the case (and in fact using evidence that doesn't necessarily support a hypothesis of mutualism over other ecological interactions, like parasitism).[6] It is likewise sometimes assumed that it is the microbiome's activity *as a community* that is relevant in such coevolution.[7] Here, too, there is some evidence of this in some cases, such as with regard to pathogen protection, but examples like this are likely exceptions rather than the rule.[8]

We briefly highlight these complications, though not to question eubiosis as the purported goal of microbiome engineering. But given the checkered history of the word's ambiguous usage and given that it seems to imply a harmonious evolved relationship between host and microbiota that is often questionable, the term eubiosis must be treated with some caution. It is best to treat the terms eubiosis and dysbiosis as stand-ins for "good for the host" and "bad for the host" without implying further (often unsupported) content about the evolved relationships between microbiota and host.[9]

Let us turn now to two detailed examples of microbiome engineering, beginning with the poster child for microbiome-mediated microbial community restoration or regeneration: the use of fecal microbial transplantation in the treatment of *Clostridium difficile* infection. *C. difficile* infection is an increasingly common cause of diarrhea, difficult to treat and often recurring. Patients showing one recurrence are prone to additional bouts of illness: clearly they are in a state of dysbiosis. Especially at risk are those who have recently received antibiotics or with inflammatory bowel disease (IBD). It is generally thought that dysbiosis resulting from such antibiotic treatment or associated with IBD is permissive of *C. difficile* growth, which is otherwise discouraged by the resident microbial community. Recurrence rates for healthcare-associated cases have been reported to be 21 percent in the United States and 9 percent of infected people die, so *C. difficile* infection is a serious problem of increasing prevalence.

Antibiotics, often used to treat infections, further deplete the resident microbiome, unfortunately encouraging reestablishment of *C. difficile* infection. Fecal microbiome transplantation has been a recognized, if often fringe, medical procedure for several conditions and in several contexts for many centuries.[10] Transfer of fecal material from healthy donors to patients with recurring *C. difficile* infection became popular in the early 1980s and has proven remarkably successful: about 95 percent of patients recover fully. Various methods of infusing the donor microbiome into the recipient are successful, and the Dutch microbiologist Willem de Vos writes of such treatment that "the nature of the microbiota (fresh or frozen), the delivery method (duodenal or colonic), or the location, origin or dietary habits of the donors, do not affect the final outcome [confirming] the early observations that the intestinal ecosystem in *C. difficile* infection patients is so disturbed that the donor microbes rapidly start occupying the available niches resulting in a normally functioning microbiota."[11]

Although fecal microbial transplantation procedures have shown success in treating some infections that would otherwise be life-threatening, these treatments come with significant risks of their own. For example, the donor may carry pathogenic bacteria (those that cause disease) and viruses that are transferred to the recipient along with the desired healthful material. So for non-life-threatening ailments, a safer alternative is needed.

One alternative which has shown some promise for treating illnesses like antibiotic-associated diarrhea or irritable bowel syndrome involves administering probiotics and prebiotics. Diet is known to have a strong influence over the gut microbiota, and probiotic/prebiotic therapy takes advantage of this. Probiotics are "live microorganisms that, when administered in adequate amounts, confer a health benefit on the host" whereas prebiotics are specific carbohydrate precursors to bacterial metabolism.[12] In other words, probiotics provide microbes and prebiotics provide their food. Both are widely available in fortified foods—foods with organisms or nutrients added to increase their presumed healthfulness—such as yogurts.

The idea that fortified foods are good for the gut is likely a very ancient idea, but it received scientific credibility at the turn of the twentieth century when the bacterium *Bifidobacterium bifidum* was discovered in fecal samples of breast-fed infants.[13] Following the work of Nobel Laureate Elie Metchnikof (1845–1916) on the hypothesized role of probiotics in limiting aging (a hypothesis he thought was confirmed by the longevity of Bulgarian populations who ate sour milk), probiotics in the form of yogurts were developed for therapeutic use and sold in pharmacies.[14] Many probiotics in current use involve the same genera of bacteria that Metchnikof thought important

(*Lactobacillus* and *Bifidobacterium*) and were originally derived from the feces or the intestinal mucus membranes of healthy humans.[15] When we eat appropriately fortified foods, these bacteria enter our gut communities and potentially remain there as constituents. What do these bacteria do?

Through the process of fermentation, gut bacteria metabolize carbohydrates, proteins, and lipids that humans ingest. Some of these compounds, such as fibers, are indigestible to humans, but can be metabolized by bacteria, creating end products such as organic acids that are useful to humans. *Lactobacillus* and *Bifidobacterium* bacteria, those found in active culture in yogurts, produce lactic acids and acetic acids, respectively. These compounds are themselves important for humans, having the effects of potentially lowering pH and discouraging the growth of pathogens.[16] But as with the fecal microbiota transplant case, it is no simple one-to-one connection between introducing one species of bacteria and beneficial human health outcomes. In fact, research over the last decade seems to show that it is not the direct products of the probiotic bacteria that matter most but, instead, the end products of other bacteria these bacteria feed.[17] The complex community of microorganisms in the gut work with each other to produce health and illness.

An everyday garden analogy may help here. Let's say we want to serve cabbage for dinner in the winter and so we attempt to grow it in the backyard in the summer. There are many factors, both abiotic (sun, rain, temperature) and biotic (aphids and slugs eat cabbage, for example), that affect the growth of cabbages. How should we proceed to encourage our cabbages to grow? One option is somewhat indirect but potentially very effective: introduce ladybugs into the garden in order to control insects like aphids which may decimate our cabbage patch. However, the effectiveness of this intervention will depend on a number of uncontrolled factors: Can the ladybugs survive in the garden? Are there too many predatory birds? Have we introduced enough ladybugs?

Introducing *Lactobacillus* and *Bifidobacterium* to encourage human health outcomes is a bit like introducing ladybugs for the purposes of producing human food: it's the indirect effects that matter. These bacteria produce end products that are themselves metabolized by other bacteria (a phenomenon known as "cross-feeding") to produce acids that are directly important for human health. One important example is the short chain fatty acid butyrate, produced by bacteria from the genera *Faecalibacterium* using the end products of other bacteria in fortified foods. In a 2019 study of nearly 1,000 people living in the Netherlands, it was shown that higher fecal levels of butyrate predicted better insulin responses, thus linking bacterial butyrate production to diseases of metabolic dysfunction, such as diabetes and obesity.[18] Fortified foods are likely indirectly important for gut health.

Although the complexity of ecological interactions in the community of gut bacteria undermines any claim that a specific bacterium introduced will provide a magic bullet solution for ill health associated with a dysbiotic gut, probiotics and prebiotics generally come with low risk and potentially high reward and are thus a useful strategy for engineering a eubiotic community. And as our knowledge of the ecological complexity of the gut increases, personalized probiotics and prebiotics may prove to be an effective and safe way to treat many modern illnesses that are non-life-threatening and that involve regenerating a eubiotic gut community following disturbances like antibiotics.

Humans and the microbes that live in our guts have a long history, evolutionarily speaking, which is not only relevant to the kinds of illnesses we currently face but also to the kinds of microbiome engineering required to provide solutions.[19] Although there is considerable taxonomic variation in the composition of gut microbes among healthy humans, the magnitude of those differences starts to shrink as we "zoom out" in the tree of life.[20] Researchers have found that the microbial contents of human guts from different populations of humans living different lifestyles are more similar to each other than they are to the guts of other vertebrates or even other great apes. This suggests that the kinds of bacteria populating the guts of different species of animals reflect deep evolutionary splits as particular microbes began to associate with the differences between gut environments of different hosts (differences in pH, oxygen levels, host-derived molecules, type of digestive organs, host immune system, diet, etc.).

One well-studied example is that of *Lactobacillus reuteri*, a gut bacterium with multiple lineages that map to specific hosts. Germ-free mice (that is, experimental mice that are devoid of *Lactobacillus* bacteria) presented with *L. reuteri* from the guts of mice, humans, pigs, and chickens show colonization that is, except in a few cases, limited to the mouse-derived *L. reuteri* strain.[21] Although *L. reuteri* may be unique in its host specificity, and many microbes may colonize a wider set of hosts, this does suggest that there may be a set of microbes common to human guts, the so-called old friends. This relationship would play out in a comparison of the phylogenetic trees of hosts (say primates) and the species or strains of bacteria associated with them (say *Lactobacillus*).[22]

A 2021 study suggests that some of these "old friends" predate the split between Neanderthals and modern humans during the early Middle Pleistocene. In 2014, at an archeological site known as El Salt, in eastern Spain, researchers discovered the oldest known human coprolites, or fossilized feces. Feces are a good way of determining the composition of the gut microbiota, so it seems an obvious next step that researchers attempted to analyze the feces-containing sediments of El Salt to determine the content of Neanderthal gut microbiome from ancient DNA.[23] The data have suggested that there is a remarkable similarity between Neanderthal microbiomes and those of modern humans. Humans share with our closest living relatives, chimpanzees, a gut microbiome that is similar at the family level, but can be distinguished at the genus level.[24] Tellingly, the El Salt findings cluster closer to humans than to chimpanzees. The butyrate producer *Faecalibacterium*, a biomarker of a healthy human gut microbiome, as discussed above, is found in these samples.

Moreover, the compositional profile of the Neanderthal microbiome tends to cluster closer to the compositional profile of current humans living agrarian and hunter-gatherer lifestyles.[25] This might not be so surprising, given assumed commonalities in diet, but the data also become part of a larger picture about the adverse effects of "western" lifestyles on the composition of our gut communities.

Since epidemiologist David Strachan's "hygiene hypothesis," which emphasized the beneficial effects of microorganisms for inhibiting the development of hay fever and asthma, there has been much scientific and popular interest in the ways that sterile western lifestyles can impact the microbes that live around us.[26] The more recent "disappearing microbiota" theory hypothesizes that in fact many "western illnesses" currently on the rise (obesity, diabetes, asthma) are due to the loss of coevolved microorganisms.[27] This loss comes about from changes to modern western styles of living: diets low in fiber, sterilized environments, and the overuse of antibiotics administered to humans directly or through industrial agriculture. As the evolutionary biologist Britt Koskella and colleagues summarize,

> Many of these environmental drivers reflect relatively recent changes in human evolutionary history. As most are associated with reducing microbiome diversity, this suggests the worrying possibility of irreversible change and/or loss of diversity of the global human microbiome pool, supported by evidence of higher microbiome diversity in uncontacted Amerindians and of reduced human microbiome diversity relative to our nearest ancestors.[28]

An epidemiological study, for example, of children attending daycare in Finland showed that antibiotics were significantly associated with the development of asthma and excess body weight and that gut bacteria diversity was diminished in children who had taken antibiotics up to *two years* prior to the analysis.[29] There might also be a ratcheting effect that can occur, leading to microbial loss over multiple generations of hosts. For example, when mice with "humanized" microbiota (the microbiota from human fecal samples) are transferred from a high fiber diet (corn, soybean, wheat, oats, alfalfa, and beets) to a low fiber diet (sucrose and corn meal), their gut diversity declines and does not fully recover even after they are placed back on a high fiber diet.[30] Furthermore, if mice are kept on a low fiber diet over multiple generations, the diversity continues to decline and cannot be reestablished through diet.

These worrying trends have suggested to some that we need secure biobanks to preserve microbial diversity as we work out what microbes matter for human health. As four leading microbiomics scientists (Maria Dominguez-Bello, Rob Knight, Jack Gilbert, and Martin Blaser) write,

> Most urgently, we need to preserve the diversity of ancestral microbes from globally diverse human populations and especially include those who have had the least exposure to urbanization. Using current technology, and under the precautionary principle (to avoid the introduction of products and processes the ultimate effects of which are unknown), it is paramount that we expand the efforts to capture and preserve the human microbiota while it still exists. This is a needed step toward restoration and could help mitigate the potential risk to human health that urbanization encompasses.[31]

Losing potentially important research objects to human destruction has long been a worry for ecologists in general: one that connects pure science to conservation. The macrobial ecologist Peter Morin phrases the urgency in the following way (quoted in full in the epigraph to this chapter): given the history of evolved complexity between humans and their gut microbiota, the current trend toward reduced microbial diversity is worrisome because it may mean that "potential knowledge will be lost forever if we allow natural communities to disappear before we learn their secrets."[32] And to go back to the opening of this chapter, this is bad for successful engineering because the complexity of nature often requires a dialogue between the world of human engineered systems and those occurring naturally.

This desire to maintain existing biodiversity, if for no other reason than that we don't know enough (and don't even know what we don't know), is complemented by a desire to take charge, to engineer more effectively using modeling, technologies coming from genomics, and the approaches of "systems biology." Such engineering approaches will become increasingly popular in biomedical applications such as fecal microbial transplantation which are expanding rapidly from *C. difficile* infection to mysterious disease entities like chronic fatigue syndrome and multiple sclerosis. A "synthetic community" put together from a limited number of separately established and cultured "pure" species or strains is more suitably presented for regulatory approval as a therapeutic than is any infusion of fecal materials from donors: in addition to their individual-

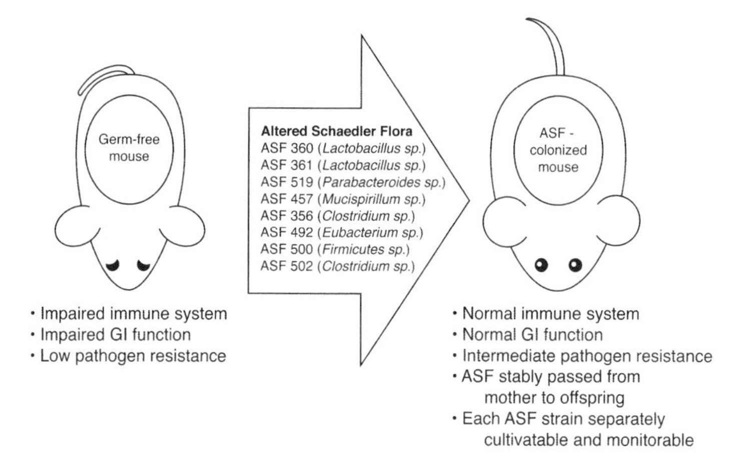

FIGURE 5.2 | Colonization of gnotobiotic mice (germ-free, born by C-section) with the eight strains of the Altered Schaedler Flora (ASF). Such colonized mice develop much more normally than do gnotobiotic mice. Each strain in the ASF can be cultured separately and relative populations of each species in any mouse can be assessed by molecular methods.

to-individual and day-to-day variability, these might carry undetectable bacterial or viral pathogens. Moreover, synthetic community approaches allow experimental additions and subtractions of individual taxa, allowing us to progress from establishing correlations to proving causation for individual species and for communities.[33]

Reconstituted synthetic communities have been around since the middle 1960s and something called the "Altered Schaedler Flora," consisting of eight independently cultivatable isolates initially obtained from the much more complex mouse gut microbiome, has been extensively used since the early 1980s, long before "microbiomics" even became a popular word (see fig. 5.2).[34] Germ-free mice inoculated with this mix have many of the properties of "normal" mice and the community is, for instance, capable of producing short-chain fatty acids to support nutritional needs of host mice. Most or all of the genomes of this eight-species mix have now been sequenced. As genomic sequencing and genome assembly from metagenomic data become even easier and cheaper, computational approaches to establishing the metabolic capabilities of individual species of the microbiome and the use of species co-occurrence data to establish patterns of cross-feeding and other interactions will drive *in silico* modeling.[35]

Modern genomic technologies offer many new approaches to synthetic community construction and testing. For instance, Paul Rainey's group at the Max Planck Institute in Germany proposed in 2019 the use of a "K-chip." This device allowed the simultaneous testing of 100,000 multispecies communities, made up of various combinations of nineteen different soil isolates, each combination in its own tiny droplet. Successful community combinations were those more supportive of the growth of a nitrogen-fixing plant symbiont.

The goal of almost all such engineering endeavors is "functional." We are aware of no synthetic community efforts intended to reassemble communities exactly as they were found before (or in their "natural" state). In fact this would be very difficult and certainly the human fecal transplantation used to treat *C. difficile* infection vary depending on the donors, even when all are healthy. Instead, the goal is to assemble a community that performs the *same function* as a preexisting and undisturbed natural one—to sing the same song, even if with different singers. We accept this as a form of "regeneration" because in fact many instances of organism-level processes taken to be regeneration involve different genes and cell types than did the original development of the missing or damaged organ or appendage (as discussed earlier). But we've replaced or at least supplemented the evolution-based meaning of "function" or "purpose" that we ascribed to microbial communities in chapter 4 with a human-design or "engineering" definition appropriate to synthetic communities. We've added a new "why." What does this mean philosophically? This and what we think we've accomplished are dealt with next in our epilogue.

Epilogue

Our goal in this book has been to offer a way of understanding how microbial community regeneration works and why it occurs. We've aspired to an understanding that is broadly applicable: to provide a framework in which the reassembly, restoration, or recurrence of a microbial community after a disturbance may be treated in the same terms as the regeneration of some or all parts of an organism after an injury. The problem was that if this were solely about the "how" (a proximate explanation), there's really nothing other than a metaphorical similarity between organismal and community regeneration. We'd want to describe the former with words appropriate to developmental biology. Even if some different genes and cell types were involved in regeneration, it's still the language of developmental biology that we'd use. Whereas for the latter, it's the language and rules of ecology. Communities come about through the assembly or recruitment of independent species, each with its own already evolved capacities and propensities.

Given our discussions throughout this book, there seem to us to be two general ways to reconnect organismal and community regeneration. Let us start with the path *not* followed in this book. This is a "systems" or cybernetic approach. The laudable new scientific paradigm of "systems biology" seeks to understand how the current functioning of a complex adaptive system reflects the integrated interaction of its several components without much attention to where such components came from or why they do what they do. Failure, according to this paradigm, does not have to do with failing to be evolutionarily adaptive but is instead captured through more abstract characterizations of the disintegration of system order or information. Similar organizational principles, suitably generalized, might govern the behavior of organisms and of microbial communities during regeneration—as systems that return to previous configurations—but these would be ahistorical principles addressing networks of interactions affecting proximate causation, not the specifics of either process in biological terms. There might be a common "how," but this would be at an abstract level above that of the actual components involved and ignorant of history or any evolutionary notion of purpose or function.

The approach we have taken, on the other hand, emphasizes seeking to find a common "why." *Why* do organisms and communities often return to former states after disturbance, and in what sense might the two types of regeneration reflect the same ultimate cause—the differential success of regenerating entities at both levels? To do that, we had to dissect the differences between ecological and evolutionary approaches to community assembly, long a difficult-to-bridge gap in the thinking of biologists.

Why-questions ask about the purpose of regenerative processes at either level. The organism often regenerates because regeneration allows that organism to have more progeny and these progeny inherit the tendency to regenerate. So traits favoring regeneration increase in frequency in a population. The problem with communities (of macrobial species *or* microbes), *qua* communities, is that mostly they don't have progeny—they don't reproduce—so there's no way for selection as normally understood to get a handle on them. This has always been the problem with inferring "ecosystem function": it is decidedly not the same thing as the "function" of the vertebrate eye or the salamander limb. This is one important difference that has long kept ecology and evolutionary biology apart as disciplines.

We suggest that the way out of this is to sidestep (most) traditional Darwinian thinking, which is (mostly) based on what's called Lewontin's Recipe (again, that natural selection involves heritable variation in fitness) and (now) multilevel selection theory, which also requires reproduction. Instead, if we adopt the currently less popular replicator/interactor framework, proposed decades ago by the philosopher David Hull, we might be able to see frequently recurring or more persistent microbial communities as harboring genes that favor the community-level interactions that make such communities more frequent or persistent. So the differential success of a community type (which Hull would call an "interactor") favors the differential replication of the genes, organisms, or species that contribute to that differential success. The community itself doesn't have to reproduce.

This solves the philosophical problem posed by the nonreproduction of microbial communities and, thus, the problem of the "why" question. Organisms sometimes possess regenerative abilities because natural selection can favor (through their differential reproduction) those organisms that can regenerate. Communities likewise sometimes possess regenerative abilities because natural selection can favor the genes (or organisms or species) that make these types of communities recur more frequently or last longer. Hull's formulation works for both situations, while Lewontin's Recipe only works for entities that reproduce.

This is only what philosophers would call a "how possibly" solution, though: it is a story about how it all *possibly* works. It was important to resolve this problem, which has led to heated debates over the holobiont concept or about Gaia: both were objected to by most Darwinists because their evolution by natural selection was "impossible in theory." The replicator/interactor formulation makes it legitimate to think about genes, organisms, or even species that promote the recurrence or persistence of particular communities (the interactors) as themselves the replicating units that are favored. Thus organismal and microbial community regeneration could, when viewed in the right light, be "the same" thing, or at least answer the "why" question in the same way.

Whether they *actually* do or not is what philosophers would call a "how actually" question, and though it remains unanswered, it does indicate a direction future empirical research could take. We'd need to show that the regenerative abilities of some communities were naturally selected for. Empirically, we could look for evidence that there is selection on genes that are frequently laterally transferred *to be* frequently transferred (increasing community resilience to disturbance by increasing redundancy in the species able to perform critical community functions). If we could show that such genes, genes highly subject to lateral gene transfer, "belong" to communities more than to organismal lineages, that would be a suggestive start.

Ironically, given that their aim is creating novel communities rather than investigating the actual ones that have evolved on this planet, efforts to build synthetic communities (reviewed in chapter 5) might hold the key to answering this "how actually" question. Also ironically, again given their aim, these efforts can support a powerful argument in favor of environmental preservation.

The key is this. There may well be genes already out there that more obviously contribute to community functions than they do to organismal ones. The goal of engineering novel communities with regenerative abilities may encourage us to find the genes most responsible for community regeneration, and by investigating them through synthetic reconstruction we may be able to establish that *actual* communities have *actually* evolved by natural selection to house them. Furthermore, to successfully build communities with species housing these genes will itself be to provide a "how actually" explanation, in a way. We will have, in such a case, created *actual* communities with regenerative abilities and possessing the engineering function of regeneration. This *is* purposive community regeneration *actualized*.

The argument in favor of environmental preservation is this.

The epigraph from Peter Morin with which we begin chapter 5 holds that a strong reason to continue investigating, and to preserve, natural systems is the "secrets" they contain. Their secrets, forever lost if they are destroyed, will include the genes responsible for community regeneration. The physicist Richard Feynman famously said, "What I cannot create, I do not understand." But similarly, what I do not understand, I cannot create. No rational engineer would ignore the secrets coming from the investigation of natural systems and no rational biologist or philosopher can afford to ignore such secrets as attempts to create synthetic communities by cobbling together existing species now might reveal. It doesn't matter whether we take a systems approach (asking only proximate questions) or an evolutionary one (seeking ultimate answers). With either approach, one of the many reasons to conserve biodiversity, not only for big organisms we can see and romanticize, like the polar bear, but for tiny invisible organisms like *B. plebeius*, is that to lose them is to lose information that we might need to rebuild future ecosystems along functional lines.

And beyond engineering, there is, lurking in the background, an even deeper argument in favor of environmental preservation. A perennial question in ethics has been about the kinds of living systems that have *moral standing*. These are systems with intrinsic (noninstrumental) value; systems whose "interests" are worthy of respect in their own right and of being fully taken into consideration before an action involving them is performed. One influential answer to this question is that moral standing requires the possession of evolutionary function, and in particular, the functions to respond to injury and to regenerate.[1] If we can show that communities, as inter-actors, possess these functions, we will have (inadvertently) taken a step in a moral direction as well. The ability to regenerate, whatever we have done to them, is a measure of their value. We do wrong when we destroy them.

Acknowledgments

Even a very short book, like this one, incurs many debts.

The book is a result of the fortunate convergence of two social-intellectual threads. The first traces back to 2016, and to Dalhousie University, where one of the coauthors (Andrew Inkpen, a historian and philosopher of biology) was a postdoctoral fellow under the other (Ford Doolittle, a molecular biologist turned philosopher). Our conversations at this time—nurtured by the wonderful people interested in biology and philosophy at Dalhousie in the Evolution Studies Group—centered on the science of metagenomics and its implications for the theory of evolution by natural selection. Metagenomics, particularly when applied to the microbial world, seemed to us to offer a new lens for reflecting on long-standing questions in biology about the evolution of groups, ones tracing back to Darwin's *Origin of Species*. This book builds on those conversations, so a warm thanks is due to everyone at Dalhousie for cultivating a wonderful and rigorous intellectual environment. Ford, and through him Andrew, was supported at Dalhousie by the National Sciences and Engineering Research Council of Canada (grant GLDSU447989), the Gordon and Betty Moore Foundation (GBMF9729, https://doi.org/10.37807/GBMF9729), and the New Frontiers in Research Fund (grant NFRFE-2019-00703).

The second thread began in 2018 at the Marine Biological Laboratory at Woods Hole, when Jane Maienschein and Kate MacCord approached Andrew (then at Brandon University) about their exciting regeneration project, funded by the James S. McDonnell Foundation and encouraged by its president, Susan Fitzpatrick. The idea of a series of volumes about regeneration that showcases what can be accomplished when we think across traditional disciplinary boundaries and across the scales of life was conceived of by Jane, Kate, and Susan, in consultation with the University of Chicago Press (special thanks to editor Joseph Calamia), and with members of the growing regeneration project team. As a result of a successful weeklong summer seminar at the Marine Biological Laboratory in 2019, Andrew came to see a strong connection between regeneration and his conversations at Dalhousie. When the time came to write, he asked Ford to join the project, and together they applied some of their earlier thinking to the task of understanding how to make sense of the regeneration and evolution of microbial communities. Leaders of the various regeneration project working groups read the manuscript in full and engaged in many helpful discussions about its logic, structure, and writing. These people include (beyond Jane and Kate) Kat Maxson Jones, Lucie Laplane, Fritz Davis, Jim Collins, and Manfred Laubichler.

Many friends, students, and colleagues not directly associated with the regeneration project team or the University of Chicago Press have read drafts of chapters of this book or listened carefully to the material presented. The book has benefited substantially from their helpful and critical feedback. These include members of an informal group of Ford's current and former postdocs (Tyler Brunet, Carlos Mariscal, Rose Novick, Celso Neto, Chris Jones, and Chris Lean) and core members of the Gaia Working Group (Celso Neto, Chris Jones, Joe Bielawski, Chris Lean, Richard Boyle, and Letitia Meynell). MacKenzie Scott, a biology student at Mount Allison University, read the text in full over the summer of 2021—because of her careful work, the text is more logically sound and more accessible for its intended audience. Archivists at the New York Botanical Garden and the American Heritage Center at the University of Wyoming deserve a special thanks for providing images for the book from their collections during a challenging year of closures.

The largest debt incurred is more personal. Andrew's son was born just as the writing for this book began in earnest, and he owes an incalculable debt to his wife Dani for her continued encouragement and support throughout a sleep-deprived year. Despite the exhaustion, Dani's insights were invaluable and her mark can be found in all of the more well-written sections of the book.

Further Reading

CHAPTER 1: REGENERATION

A wonderful text introducing the idea of regeneration in biology is Richard Goss, *Principles of Regeneration* (1969). The set of essays collected in Charles Dinsmore's *A History of Regeneration Research* (2007) together provide a great historical overview of the scientific study of regeneration. The proximate-ultimate distinction, which we draw on throughout this book, has been the focus of much controversy in philosophy of biology. Although it differs from the way we use the distinction, Mayr's original discussion in "Cause and Effect in Biology" (1961) is thought provoking and worth reading. An informative historical and philosophical study of the different uses to which Mayr put the distinction is John Beatty, "The Proximate/Ultimate Distinction in the Multiple Careers of Ernst Mayr" (1994). For an evaluation of contemporary uses of Mayr's distinction, see Laland et al., "Cause and Effect in Biology Revisited: Is Mayr's Proximate/Ultimate Dichotomy Still Useful?" (2011). A great introduction to the history of microbiology, current microbiology, and the philosophy of science is Maureen O'Malley, *Philosophy of Micro-*

biology (2014). And for a discussion of debates surrounding how to define an ecological community, see Kim Sterelny, "Local Ecological Communities" (2006), Jay Odenbaugh, "Seeing the Forest *and* the Trees" (2007), and Christopher Lean, "Indexically Structured Ecological Communities" (2018).

CHAPTER 2: ECOLOGY

There is much written on the early history of ecology. For a broad overview of the history of ecological thinking, see Donald Worster, *Nature's Economy* (1994). Sharon Kingsland, *The Evolution of American Ecology* (2005), and Robert Kohler, *Landscapes and Labscapes* (2002), both provide detailed analyses of ecology in the early twentieth century. The most detailed account of debates between Gleason and Clements is Christopher Eliot, "The Legend of Order and Chaos" (2011). A great introduction to community ecology, particularly theories of succession and community assembly, is Peter Morin, *Community Ecology* (2011). Mark Vellend, whose work we summarize above, provides a more detailed discussion of his framework in his *The Theory of Ecological Communities* (2016). Two great and very readable scientific papers applying ecological and evolutionary theory to the gut microbiome are Costello et al., "The Application of Ecological Theory Toward an Understanding of the Human Microbiome" (2012), and Foster et al., "The Evolution of the Host Microbiome as an Ecosystem on a Leash" (2017). Helpful introductions to microbial ecology can be found in chapter 6 of Angela Douglas, *Fundamentals of Microbiome Science* (2018), and chapter 5 of Maureen O'Malley, *Philosophy of Microbiology* (2014).

CHAPTER 3: EVOLUTION

Good introductions to current thinking about evolution by natural selection are Peter Godfrey-Smith's books, *Darwinian Populations and Natural Selection* (2009) and (more succinctly) *The Philosophy of Biology* (2013). "Lewontin's Recipe," as first articulated by Richard Lewontin himself, can be found in Lewontin, "The Units of Selection" (1970). The literature on "holobionts" includes a 2016 book, *The Holobiont Imperative* (2016) by Thomas Bosch and David Miller, and many papers, notably Eugene Rosenberg and Ilana Zilber-Rosenberg, "The Hologenome Concept of Evolution after 10 Years" (2018). Michael Ruse's *The Gaia Hypothesis: Science on a Pagan Planet* (2013) is a good summary of the Darwinian opposition, and the hypothesis itself first appeared in popular form in James Lovelock, *Gaia: A New Look at Life on Earth* (1979). A good summary, from a Gaian perspective, of the history of life on Earth is Tim Lenton and Andrew Watson, *Revolutions That Made the Earth* (2013). The ideas behind MLS1 and MLS2 are articulated, rather technically, in Samir Okasha, *Evolution and the Levels of Selection* (2006).

CHAPTER 4: INTERACTORS

Five scientific papers useful for understanding and further exploring the themes in this chapter are: Peter Turnbaugh et al., "A Core Gut Microbiome in Obese and Lean Twins" (2009), which introduces ecological thinking to microbiomics; Stilianos Louca et al., "Function and Functional Redundancy in Microbial Systems" (2018), which highlights a functional definition of "community"; David Hull,

"Individuality and Selection" (1980), is foundational to our approach here, in particular his articulation of the idea of "interactors"; Laurent Philippot et al., "Microbial Community Resilience across Ecosystems and Multiple Disturbances" (2021), which provides a recent summary of the state of the art; and our own "Processes and Patterns of Interaction as Units of Selection: An Introduction to ITSNTS Thinking" (2018) expands on a theory alluded to here. Useful books include David Quammen, *The Tangled Tree: A Radical New History of Life* (2018), which reviews the development of the current Tree of Life and the role of lateral gene transfer; Richard Dawkins, *The Selfish Gene* (1976), is foundational to much evolutionary thought; and Andrew Hendry, *Eco-Evolutionary Dynamics* (2009), provides a good review of community and ecosystem evolution.

CHAPTER 5: ENGINEERING

For popular introductions to the gut microbiome, microbiome engineering, and the problems with modern diets and the overuse of antibiotics, see Martin Blaser, *Missing Microbes* (2014), Ed Yong, *I Contain Multitudes* (2016), and Justin and Erica Sonnenburg, *The Good Gut* (2015). A great and very readable textbook introduction to how microbes shape animal biology in general is Angela Douglas, *Fundamentals of Microbiome Science* (2018). The latter book also discusses the history and coevolution of microbes and hosts, including early hominids. Eugene Rosenberg and Ilana Zilber-Rosenberg, *The Hologenome Concept* (2014), as well as Eugene Rosenberg, *Microbiomes* (2021), provide excellent sources for understanding holobionts and the hologenome idea.

Notes

CHAPTER 1

1. R. Goss, *Principles of Regeneration* (New York: Academic Press, 1969), 1.
2. Charles E. Dinsmore, *A History of Regeneration Research: Milestones in the Evolution of a Science* (Cambridge: Cambridge University Press, 2007); Jane Maienschein and Kate MacCord, *What Is Regeneration?* (Chicago: University of Chicago Press, 2022).
3. Daniel Ogden, *Drakon: Dragon Myth and Serpent Cult in the Greek and Roman Worlds* (Oxford: Oxford University Press, 2013).
4. Goss, *Principles of Regeneration.*
5. Maienschein and MacCord, *What Is Regeneration?*
6. A separate terminological issue (and possibly also a conceptual issue) exists in the background here about the difference between regeneration and generation. Distinctions between regeneration, generation, growth, and asexual reproduction might be a point of some contention, even in cases of individual regeneration. When you cut a hydra into pieces and each piece makes a new hydra, was that regeneration or asexual reproduction? Perhaps such distinctions start to break down at the extremes. For communities, which do not "develop" in the developmental biology sense, it is hard to know what would amount to a distinction between generation and regeneration. As we note in this chapter, Frederic Clements called regeneration only what occurs when an established community is disturbed in some way (he thought "secondary succession" was regeneration and "primary succession" was generation). We allow for a broader idea of regeneration to include also when an entire community is destroyed and

a new community develops to take its place. This fits with the way ecologists currently use the term regeneration. This might collapse regeneration and generation, but we would be fine with that. We thank an anonymous reviewer for forcing us to grapple with these issues.

7. "regeneration, n." *Oxford English Dictionary*, online ed., accessed June 14, 2021, https://www.oed.com/view/Entry/161223?redirectedFrom=regeneration.

8. Clements, like others at this time, thought there were three types of individual organisms: the individual cell, the "community of cells" or multicellular organism, and the "community of organisms" or simply community. His colleague W. B. McDougall wrote in his *Plant Ecology* (1927), "Just as the second type of individual is a community composed of individuals of the first type, so a plant community, which we consider as the third type of individual, is composed of individuals of the second type, largely, though some of the component individuals may be of the first type. It is a little difficult at first to think of a plant community as an individual in the sense that a tree is an individual, because we have not been in the habit of so considering it, but there is no greater degree of difference between a tree and a plant community, such as a forest, than there is between a tree and a one-celled alga, and we shall find that the forest community has a life cycle and can do practically everything that the tree can do." (W. B. McDougall, *Plant Ecology*, 1927, 208, http://archive.org/details/in.ernet.dli.2015.271691.)

9. Frank Edwin Egler, *The Nature of Vegetation, Its Management and Mismanagement: An Introduction to Vegetation Science* (Norfolk, CT: Aton Forest, 1977).

10. That microorganisms are also good at "horizontal" or "lateral" gene transfer (see chap. 4), exchanging functional genetic material between organisms so different that even skeptical microbiologists consider them different species, is another problem.

11. There is a deep and philosophically interesting rabbit hole here which we avoid. That is the question of how we define taxonomy and function for microbial communities. For example, because both taxonomy and function can be divided up differently, this will affect our classification of communities and our judgments about their diversity. If we choose coarse grains of analysis for taxonomy (phylum) and function (metabolism), we will get less diversity than if we choose finer grains (species and carbohydrate metabolism at a particular pH and temperature, respectively). See a discussion of this issue in S. Andrew Inkpen et al., "The Coupling of Taxonomy and Function in Microbiomes," *Biology & Philosophy* 32, no. 6 (December 2017): 1225–43, https://doi.org/10.1007/s10539-017-9602-2.

12. Paul A. Oliphint et al., "Regenerated Synapses in Lamprey Spinal Cord Are Sparse and Small Even after Functional Recovery from Injury," *Journal of Comparative Neurology* 518, no. 14 (2010): 2854–72.
13. Goss, *Principles of Regeneration*, 3.
14. Angela E. Douglas, *Fundamentals of Microbiome Science: How Microbes Shape Animal Biology* (Princeton, NJ: Princeton University Press, 2018).
15. Maureen O'Malley, *Philosophy of Microbiology* (Cambridge: Cambridge University Press, 2014).
16. Ricardo Cavicchioli et al., "Scientists' Warning to Humanity: Microorganisms and Climate Change," *Nature Reviews Microbiology* 17, no. 9 (September 2019): 569, https://doi.org/10.1038/s41579-019-0222-5.
17. Philipp Engel et al., "The Bee Microbiome: Impact on Bee Health and Model for Evolution and Ecology of Host-Microbe Interactions," *MBio* 7, no. 2 (May 4, 2016): e02164-15. https://doi.org/10.1128/mBio.02164-15.
18. Cassondra L. Vernier et al., "The Gut Microbiome Defines Social Group Membership in Honey Bee Colonies," *Science Advances* 6, no. 42 (October 1, 2020): eabd3431, https://doi.org/10.1126/sciadv.abd3431.
19. J. Lederberg, "Infectious History," *Science* 288, no. 5464 (April 14, 2000): 287–93, https://doi.org/10.1126/science.288.5464.287.
20. Margaret McFall-Ngai et al., "Animals in a Bacterial World, a New Imperative for the Life Sciences," *Proceedings of the National Academy of Sciences* 110, no. 9 (February 26, 2013): 3229–36, https://doi.org/10.1073/pnas.1218525110.
21. Catherine A. Lozupone et al., "Diversity, Stability and Resilience of the Human Gut Microbiota," *Nature* 489, no. 7415 (September 2012): 220–30, https://doi.org/10.1038/nature11550.
22. Martin Blaser, *Missing Microbes: How the Overuse of Antibiotics Is Fueling Our Modern Plagues* (Toronto: HarperCollins, 2014).
23. Ernst Mayr, "Cause and Effect in Biology," *Science* 134, no. 3489 (1961): 1502.
24. Although most people agree the distinction is important, there has been considerable disagreement among philosophers of biology about how best to understand the proximate-ultimate distinction. Even Mayr's own use of the distinction changed over time. We use this distinction in a specific way that is helpful for our purposes in this book and explain our meaning below, but this differs slightly from the way in which some others use this distinction and even with some aspects of how Mayr understood the distinction in 1961 (see n. 25). To learn more about this distinction and its history, see the suggestions given in "Further Reading" at the end of this book.
25. This Aristotelian distinction, made popular for biologists by one of its

more philosophically minded practitioners, Ernst Mayr, but now considered philosophically problematic, means the following to us. *Ultimate* or *final* causes are evolutionary: the answers to questions about *why* structures or processes are present in organisms or communities involve evolution by natural selection. *Proximate* causes, for us, are the immediate activities of genes and their products, or the behaviors or characteristics of organisms or species that answer questions about *how* something (regeneration in this case) is brought about in organisms or communities.

26. Maienschein and MacCord, *What Is Regeneration?*

CHAPTER 2

1. David A. Perry, *Forest Ecosystems* (Baltimore, MD: Johns Hopkins University Press, 1994).
2. Peter J. Morin, *Community Ecology*, 2nd ed. (Hoboken, NJ: John Wiley & Sons, 2011).
3. Frederic E. Clements, *Plant Succession; An Analysis of the Development of Vegetation*, Publication no. 242 (Washington, DC: Carnegie Institution of Washington, 1916).
4. Clements, *Plant Succession*, 6.
5. H. A. Gleason, "The Individualistic Concept of the Plant Association," *Bulletin of the Torrey Botanical Club* 53, no. 1 (1926): 7–26, https://doi.org/10.2307/2479933.
6. Gleason, "Individualistic Concept," 13.
7. Gleason, "Individualistic Concept," 14.
8. Gleason, "Individualistic Concept," 16.
9. John H. Lawton, "Are There General Laws in Ecology?," *Oikos* 84, no. 2 (1999): 177–92; Stefan Linquist, "Against Lawton's Contingency Thesis; or, Why the Reported Demise of Community Ecology Is Greatly Exaggerated," *Philosophy of Science* 82, no. 5 (2015): 1104–15; Robert P. McIntosh, "The Background and Some Current Problems of Theoretical Ecology," *Synthese* 43, no. 2 (1980): 195–255.
10. Mark Vellend, "Conceptual Synthesis in Community Ecology," *Quarterly Review of Biology* 85, no. 2 (June 2010): 183, https://doi.org/10.1086/652373. See also Mark Vellend, *The Theory of Ecological Communities* (Princeton, NJ: Princeton University Press, 2016).
11. Lozupone et al., "Diversity, Stability and Resilience of the Human Gut Microbiota."

12. In chapter 5 we will provide some examples, building on these processes, of how we actually attempt to engineer this regeneration.
13. Maureen A. O'Malley, "The Nineteenth Century Roots of 'Everything Is Everywhere,'" *Nature Reviews Microbiology* 5, no. 8 (August 2007): 647–51, https://doi.org/10.1038/nrmicro1711; Maureen A. O'Malley, "'Everything Is Everywhere: But the Environment Selects': Ubiquitous Distribution and Ecological Determinism in Microbial Biogeography," *Studies in History and Philosophy of Science Part C: Studies in History and Philosophy of Biological and Biomedical Sciences* 39, no. 3 (September 2008): 314–25, https://doi.org/10.1016/j.shpsc.2008.06.005.
14. Blaser, *Missing Microbes.*
15. J. B. Losos, "Ecological Character Displacement and the Study of Adaptation," *Proceedings of the National Academy of Sciences* 97, no. 11 (May 23, 2000): 5693–95, https://doi.org/10.1073/pnas.97.11.5693; Yoel E. Stuart et al., "Character Displacement Is a Pattern: So, What Causes It?," *Biological Journal of the Linnean Society* 121, no. 3 (July 2017): 711–15, https://doi.org/10.1093/biolinnean/blx013.
16. Note, "lateral" gene transfer and "horizontal" gene transfer are two names for the same process, involving at most a segment of a chromosome. Lateral gene transfer is described in much greater detail in chap. 4.
17. Martin J. Blaser, "Our Missing Microbes: Short-Term Antibiotic Courses Have Long-Term Consequences," *Cleveland Clinic Journal of Medicine* 85, no. 12 (December 2018): 928–30, https://doi.org/10.3949/ccjm.85gr.18005.
18. Scott F. Gilbert, Jan Sapp, and Alfred I. Tauber, "A Symbiotic View of Life: We Have Never Been Individuals," *Quarterly Review of Biology* 87, no. 4 (2012): 325–41.

CHAPTER 3

1. Peter Godfrey-Smith, "Darwinism and Cultural Change," *Philosophical Transactions of the Royal Society B: Biological Sciences* 367, no. 1599 (August 5, 2012): 2161, https://doi.org/10.1098/rstb.2012.0118.
2. Richard Levins and Richard Lewontin, *The Dialectical Biologist* (Cambridge, MA: Harvard University Press, 1985).
3. Kevin Laland, John Odling-Smee, and John Endler, "Niche Construction, Sources of Selection and Trait Coevolution," *Interface Focus* 7, no. 5 (2017): 20160147.

4. Britt Koskella, Lindsay J. Hall, and C. Jessica E. Metcalf, "The Microbiome beyond the Horizon of Ecological and Evolutionary Theory," *Nature Ecology & Evolution* 1, no. 11 (November 2017): 1606–15, https://doi.org/10.1038/s41559-017-0340-2; Ross S. McInnes et al., "Horizontal Transfer of Antibiotic Resistance Genes in the Human Gut Microbiome," *Current Opinion in Microbiology* 53 (February 1, 2020): 35–43, https://doi.org/10.1016/j.mib.2020.02.002.
5. Godfrey-Smith, "Darwinism and Cultural Change," 2161; emphasis ours.
6. W. Ford Doolittle and Carmen Sapienza, "Selfish Genes, the Phenotype Paradigm and Genome Evolution," *Nature* 284, no. 5757 (1980): 601–3; Leslie E. Orgel and Francis H. C. Crick, "Selfish DNA: The Ultimate Parasite," *Nature* 284, no. 5757 (1980): 604–7.
7. David Jablonski, "Developmental Bias, Macroevolution, and the Fossil Record," *Evolution and Development* 22 (2020): 103–25, https://doi.org/10.1111/ede.12313; David Jablonski, "Scale and Hierarchy in Macroevolution," *Palaeontology* 50 (2007): 87–109, https://doi.org/10.1111/j.1475-4983.2006.00615.x.
8. Jacobus J. Boomsma, "Fifty Years of Illumination about the Natural Levels of Adaptation," *Current Biology* 26, no. 24 (2016): R1250–55.
9. Eddie K. H. Ho and Aneil F. Agrawal, "Aging Asexual Lineages and the Evolutionary Maintenance of Sex," *Evolution* 71, no. 7 (2017): 1865–75, https://doi.org/10.1111/evo.13260.
10. Camilo Mora et al., "How Many Species Are There on Earth and in the Ocean?," *PLoS Biology* 9, no. 8 (2011): e1001127.
11. Dunja Knapp et al., "Comparative Transcriptional Profiling of the Axolotl Limb Identifies a Tripartite Regeneration-Specific Gene Program," *PLoS One* 8, no. 5 (2013): e61352; M. Natalia Vergara, George Tsissios, and Katia Del Rio-Tsonis, "Lens Regeneration: A Historical Perspective," *International Journal of Developmental Biology* 62, no. 6–7–8 (2018): 351–61, https://doi.org/10.1387/ijdb.180084nv.
12. Paul G. Falkowski, Tom Fenchel, and Edward F. Delong, "The Microbial Engines That Drive Earth's Biogeochemical Cycles," *Science* 320, no. 5879 (2008): 1034–39.
13. Lynn Margulis, "Words as Battle Cries: Symbiogenesis and the New Field of Endocytobiology," *Bioscience* 40, no. 9 (1990): 673–77.
14. Ilana Zilber-Rosenberg and Eugene Rosenberg, "Role of Microorganisms in the Evolution of Animals and Plants: The Hologenome Theory of Evolution," *FEMS Microbiology Reviews* 32, no. 5 (2008): 723–35.
15. Kerry M. Oliver et al., "Facultative Symbionts in Aphids and the Horizontal Transfer of Ecologically Important Traits," *Annual Review of Entomol-*

ogy 55, no. 1 (2010): 247–66, https://doi.org/10.1146/annurev-ento-112408-085305.

16. C. A. Nalepa, D. E. Bignell, and C. Bandi, "Detritivory, Coprophagy, and the Evolution of Digestive Mutualisms in Dictyoptera," *Insectes Sociaux* 48, no. 3 (2001): 194–201.
17. "Horizontal inheritance" describes the acquisition of entire microbial cells (and all of their genes) from an environmental pool.
18. Spencer V. Nyholm and Margaret McFall-Ngai, "The Winnowing: Establishing the Squid-Vibrio Symbiosis," *Nature Reviews Microbiology* 2, no. 8 (2004): 632–42.
19. Michael S. Wollenberg and Edward G. Ruby, "Phylogeny and Fitness of *Vibrio fischeri* from the Light Organs of *Euprymna scolopes* in Two Oahu, Hawaii Populations," *ISME Journal* 6, no. 2 (2012): 352–62.
20. Kevin R. Theis et al., "Getting the Hologenome Concept Right: An Eco-Evolutionary Framework for Hosts and Their Microbiomes," *Msystems* 1, no. 2 (2016): 4.
21. In the *Origin of Species*, Darwin famously concludes, "It is interesting to contemplate an entangled bank, clothed with many plants of many kinds, with birds singing on the bushes, with various insects flitting about, and with worms crawling through the damp earth, and to reflect that these elaborately constructed forms, so different from each other, and dependent on each other in so complex a manner, have all been produced by laws acting around us. [. . .] Thus, from the war of nature, from famine and death, the most exalted object which we are capable of conceiving, namely, the production of the higher animals, directly follows. There is grandeur in this view of life, with its several powers, having been originally breathed into a few forms or into one; and that, whilst this planet has gone cycling on according to the fixed law of gravity, from so simple a beginning endless forms most beautiful and most wonderful have been, and are being, evolved." Charles Darwin, *On the Origin of Species: A Facsimile of the First Edition* (Cambridge, MA: Harvard University Press, 1964), 489–90.
22. M. Medina and J. L. Sachs, "Symbiont Genomics, Our New Tangled Bank," *Genomics* 95, no. 3 (2010): 129–37.
23. Angela E. Douglas and John H. Werren, "Holes in the Hologenome: Why Host-Microbe Symbioses Are Not Holobionts," *MBio* 7, no. 2 (2016): e02099-15, https://doi.org/10.1128/mBio.02099-15; Nancy A. Moran and Daniel B. Sloan, "The Hologenome Concept: Helpful or Hollow?," *PLoS Biology* 13, no. 12 (2015): e1002311.
24. Austin Booth, "Symbiosis, Selection, and Individuality," *Biology & Philosophy* 29, no. 5 (2014): 657–73; W. Ford Doolittle and Austin Booth, "It's

the Song, Not the Singer: An Exploration of Holobiosis and Evolutionary Theory," *Biology & Philosophy* 32, no. 1 (January 2017): 5–24, https://doi.org/10.1007/s10539-016-9542-2.

25. Moran and Sloan, "The Hologenome Concept," 5.
26. Moran and Sloan, "The Hologenome Concept."
27. There is considerable literature to the effect that there is "community heredity"—heritability at the community level (see Guilhem Doulcier et al., "Eco-Evolutionary Dynamics of Nested Darwinian Populations and the Emergence of Community-Level Heredity," *Elife* 9 [2020]: e53433; Charles J. Goodnight, "Heritability at the Ecosystem Level," *Proceedings of the National Academy of Sciences* 97, no. 17 [2000]: 9365–66). This is undoubtedly correct, but in all cases of which we are aware, something like MLS2 has been computationally (as in a model) or experimentally enforced. So that the second component in Lewontin's Recipe (see above) is found should not surprise us: it is guaranteed by the heritable properties of constituent species and there will inevitably be community-level properties determined by those species on which selection can act. Indeed this is what current modeling studies show: that some degree of enforced or natural "vertical inheritance" is required to convert the coevolutionary process into selection at a collective level (see Joan Roughgarden et al., "Holobionts as Units of Selection and a Model of Their Population Dynamics and Evolution," *Biological Theory* 13, no. 1 [2018]: 44–65; Simon Van Vliet and Michael Doebeli, "The Role of Multilevel Selection in Host Microbiome Evolution," *Proceedings of the National Academy of Sciences* 116, no. 41 [2019]: 20591–97).
28. Sébastien Dutreuil, "James Lovelock's Gaia Hypothesis: 'A New Look at Life on Earth' . . . for the Life and the Earth Sciences," in *Dreamers, Visionaries, and Revolutionaries in the Life Sciences*, edited by Oren Harman and Michael R. Dietrich, 272–87 (Chicago: University of Chicago Press, 2018); Michael Ruse, *The Gaia Hypothesis: Science on a Pagan Planet* (Chicago: University of Chicago Press, 2013).
29. James E. Lovelock and Lynn Margulis, "Atmospheric Homeostasis by and for the Biosphere: The Gaia Hypothesis," *Tellus* 26, no. 1–2 (1974): 2–10.
30. W. Ford Doolittle, "Is Nature Really Motherly?," *CoEvolution Quarterly* 29 (1981): 58–63.
31. Doolittle, "Is Nature Really Motherly?," 60–61.
32. Richard Dawkins, *The Extended Phenotype: The Long Reach of the Gene* (New York: Oxford University Press, 2016).
33. Peter Godfrey-Smith, "The Ant and the Steam Engine," review of James Lovelock, *A Rough Ride to the Future*, *London Review of Books* 37, no. 4

(February 19, 2015), https://www.lrb.co.uk/the-paper/v37/n04/peter-godfrey-smith/the-ant-and-the-steam-engine?referrer=https%3A%2F%2Fwww.google.com%2F.

34. Will Steffen et al., "The Emergence and Evolution of Earth System Science," *Nature Reviews Earth & Environment* 1, no. 1 (2020): 54–63.

35. Stephen Henry Schneider and Randi Londer, *Coevolution of Climate and Life* (San Francisco: Sierra Club Books, 1984).

CHAPTER 4

1. Maienschein and MacCord, *What Is Regeneration?*

2. Peter J. Turnbaugh et al., "A Core Gut Microbiome in Obese and Lean Twins," *Nature* 457, no. 7228 (January 2009): 480–84, https://doi.org/10.1038/nature07540.

3. Turnbaugh et al., "Core Gut Microbiome," 483–84.

4. As to *recruitment* of organisms of a species into a community, recall the mechanisms implied in figure 3.1. On the left-hand side of that diagram, we imagine that all communities dissolve and release the organisms of whatever species they contain into a common pool. The next generation of communities is randomly "recruited" from this pool and there's really no way to say that any particular community in any one generation is the "parent" of any in the next generation: there's no reproduction at this level. Nevertheless, if communities with some particular combination of species (A, B, and C in fig. 3.1) grow larger (encourage the reproduction of members of species A, B, and C better) than others, future generations of communities will increasingly come to resemble them.

5. Jonathan N. V. Martinson et al., "Rethinking Gut Microbiome Residency and the Enterobacteriaceae in Healthy Human Adults," *ISME Journal* 13, no. 9 (September 2019): 2306–18, https://doi.org/10.1038/s41396-019-0435-7.

6. Stilianos Louca et al., "Function and Functional Redundancy in Microbial Systems," *Nature Ecology & Evolution* 2, no. 6 (June 2018): 936–43, https://doi.org/10.1038/s41559-018-0519-1.

7. Morgan G. I. Langille et al., "Predictive Functional Profiling of Microbial Communities Using 16S RRNA Marker Gene Sequences," *Nature Biotechnology* 31, no. 9 (2013): 814–21.

8. Jeffrey G. Lawrence and Howard Ochman, "Amelioration of Bacterial Genomes: Rates of Change and Exchange," *Journal of Molecular Evolution* 44, no. 4 (April 1, 1997): 383–97, https://doi.org/10.1007/PL00006158.

9. Miriam Land et al., "Insights from 20 Years of Bacterial Genome Sequencing," *Functional & Integrative Genomics* 15, no. 2 (2015): 141–61.
10. Jan-Hendrik Hehemann et al., "Transfer of Carbohydrate-Active Enzymes from Marine Bacteria to Japanese Gut Microbiota," *Nature* 464, no. 7290 (2010): 908–12.
11. Jan-Hendrik Hehemann et al., "Bacteria of the Human Gut Microbiome Catabolize Red Seaweed Glycans with Carbohydrate-Active Enzyme Updates from Extrinsic Microbes," *Proceedings of the National Academy of Sciences* 109, no. 48 (2012): 19786–91.
12. Laura Eme et al., "Lateral Gene Transfer in the Adaptation of the Anaerobic Parasite Blastocystis to the Gut," *Current Biology* 27, no. 6 (March 20, 2017): 807–20, https://doi.org/10.1016/j.cub.2017.02.003.
13. George Karam et al., "Antibiotic Strategies in the Era of Multidrug Resistance," *Critical Care* 20, no. 1 (2016): 1–9.
14. P. J. Hastings, Susan M. Rosenberg, and Andrew Slack, "Antibiotic-Induced Lateral Transfer of Antibiotic Resistance," *Trends in Microbiology* 12, no. 9 (September 1, 2004): 401–4, https://doi.org/10.1016/j.tim.2004.07.003.
15. Jeffrey G. Lawrence and John R. Roth, "Selfish Operons: Horizontal Transfer May Drive the Evolution of Gene Clusters," *Genetics* 143, no. 4 (1996): 1843–60.
16. Goss, *Principles of Regeneration*, 3.
17. Richard Dawkins, *The Selfish Gene* (New York: Oxford University Press, 1976), 21.
18. David L. Hull, "Individuality and Selection," *Annual Review of Ecology and Systematics* 11 (1980): 318.
19. Roughgarden et al., "Holobionts as Units of Selection."
20. Thomas G. Whitham et al., "A Framework for Community and Ecosystem Genetics: From Genes to Ecosystems," *Nature Reviews Genetics* 7, no. 7 (July 2006): 510–23, https://doi.org/10.1038/nrg1877; Thomas G. Whitham et al., "Intraspecific Genetic Variation and Species Interactions Contribute to Community Evolution," *Annual Review of Ecology, Evolution, and Systematics* 51, no. 1 (2020): 587–612, https://doi.org/10.1146/annurev-ecolsys-011720-123655.
21. Whitham et al., "Intraspecific Genetic Variation," 588.
22. In ecology as a science, a dispute still rages between those who hold that ecological niches are real and (often) finely defined and those (so-called neutralists) who think that species assemble mostly on the basis of who gets there first (see chap. 2). Still, there is a commitment to the notion of trophic levels, broadly defined, and a widespread faith that if sunlight favors the evolution or recruitment of photosynthesizers, herbivores will

arise or be recruited to eat them, and carnivores to eat *them*. At some level, however coarse, such "evolutionarily stable states," as they are called, will recur, and their recurrence does differentially perpetuate the replicators (often genes) that underwrite their formation.

23. Peter Godfrey-Smith, *Philosophy of Biology* (Princeton, NJ: Princeton University Press, 2014), 79.
24. Laurent Philippot, Bryan S. Griffiths, and Silke Langenheder, "Microbial Community Resilience across Ecosystems and Multiple Disturbances," *Microbiology and Molecular Biology Reviews* 85, no. 2 (2021): e00026-20.

CHAPTER 5

1. Morin, *Community Ecology*, 348.
2. Actually, engineering-type and evolutionary-type ultimate explanations *might* both be appropriate for this system, given a broad enough understanding of evolutionary-type ultimate explanations. Bicycle seats are products of both advance designing by bicycle engineers and "selective" purchases by bicycle riders. That's a story we will not pursue here.
3. Justin L. Sonnenburg, "Microbiome Engineering," *Nature* 518, no. 7540 (February 26, 2015): S10. https://doi.org/10.1038/518S10a.
4. Noel T. Mueller et al., "The Infant Microbiome Development: Mom Matters," *Trends in Molecular Medicine* 21, no. 2 (2015): 109–17.
5. Maria G. Dominguez-Bello et al., "Partial Restoration of the Microbiota of Cesarean-Born Infants via Vaginal Microbial Transfer," *Nature Medicine* 22, no. 3 (2016): 250.
6. It is true that the species within a microbiome have often "coevolved" with their hosts, but this does not entail mutualism. One way to show this is through the use of phylogenetic trees. Every time the host species (call it species A) splits into two species (B and C), an associated microbial species also splits, so phylogenetic ("family") trees of the host and of the microbe will have the same overall structure (see Robert M. Brucker and Seth R. Bordenstein, "The Hologenomic Basis of Speciation: Gut Bacteria Cause Hybrid Lethality in the Genus Nasonia," *Science* 341, no. 6146 [2013]: 667–69). But parasites and predators also coevolve with their hosts and often "cospeciate" with them (see Douglas and Werren, "Holes in the Hologenome"). So something like a "phylosymbiotic" pattern in which a phylogenetic analysis of a microbe is congruent with an evolutionary tree including the host is observed, but few would call the relationship of parasites and predators with their hosts and prey "mutually beneficial."

Others note that, regardless of whether such relationships are harmful, harmless, or beneficial, the term coevolution means only that species A evolves as if species B were part of its environment (Moran and Sloan, "The Hologenome Concept"). The current global pandemic is seeing selection for more infectious variants of SARS-CoV-2, after all, but this is a good thing only from the perspective of the virus. Popular use of the word "co-evolution" often implies more cooperative behavior than the evolutionary process properly described by it necessarily entails. Even coevolved mutualistic relationships, in which members of species A can't survive without members of species B and vice versa does not mean that an individual of A *plus* an individual of B make up an emergent higher level Darwinian individual in the sense in which traditional Darwinian theory recognizes individuals. Such mutualistic relationships involving multiple species do not require that those species reproduce in concert (that they show "vertical inheritance," as in chap. 3), so only a replicator/interactor framework as discussed in the chapter 4 can handle them.

7. As to the roles of the so-called pathogens, a good example not related to human biology comes from a 2018 study involving the common experimental plant *Arabidopsis* and one of its pathogens, the bacterium *Pseudomonas syringae* (see Maureen Berg and Britt Koskella, "Nutrient-and Dose-Dependent Microbiome-Mediated Protection against a Plant Pathogen," *Current Biology* 28, no. 15 [2018]: 2487–92). Investigators showed that spraying plants with microbial mixtures like the communities normally associated with them provided protection against infection and concluded that "microbiome-mediated protection is most likely not caused by one single microbial species or strain but rather by the presence of the community itself" (Berg and Koskella, "Nutrient-and Dose-Dependent Microbiome-Mediated Protection against a Plant Pathogen," 2490–91).

8. And there are other issues besides. Some dissenters have argued that the holobiont-inspired community view, and with it the concepts of eubiosis and dysbiosis, add very little to already established understandings of the relationships of microbes to hosts. These are best evaluated on an individual (microbial) species to (host) species basis: there is no need for community thinking, such skeptics hold. For example, the microbial genome analyst Katarzyna Hooks and the philosopher Maureen O'Malley have provided an extensive historical analysis of the term dysbiosis and a survey of its contemporary use. Interestingly, they "found three main uses of dysbiosis: as general change in the microbiota composition (e.g., alteration, perturbation, abnormal composition, and loss of diversity), as an imbalance in composition (almost always deemed to have negative effects),

and as changes to specific lineages in that composition (any named taxon change)" (Katarzyna B. Hooks and Maureen A. O'Malley, "Dysbiosis and Its Discontents," *MBio* 8, no. 5 [November 8, 2017]: 17, https://doi.org/10.1128/mBio.01492-17). Often these uses are vaguely defined. "Imbalance" is a problematic concept and often mere correlation is conflated with causation: what might be a response of the microbiota to the host's disease is taken to be that disease's cause. Similarly, ecological concepts now seen as controversial, such as that biodiversity implies stability are uncritically imported by microbiologists with little current knowledge of the ecological literature (see Frank Pennekamp et al., "Biodiversity Increases and Decreases Ecosystem Stability," *Nature* 563, no. 7729 [2018]: 109–12). In addition to justifying a return to more traditional clinical microbiology thinking, in which specific "pathogens" cause specific diseases, it is possible that a more function-focused approach to dysbiosis will produce better results. It is certainly the "restoration" of microbiome *function* that is the goal of many therapeutic interventions.

9. S. Andrew Inkpen, "Health, Ecology and the Microbiome," *ELife* 8 (April 17, 2019): e47626, https://doi.org/10.7554/eLife.47626.
10. It is also a prime example of successful intervention, restoration, or possibly "regeneration" of a protective microbial community. As Willem de Vos reviews this literature, prominent and ongoing veterinary practices include feeding calves on the cuds brought up by mature cows to encourage establishment of efficient rumen microbiomes, and oral inoculation of chicks with fecal preparations from adult chickens as a preventative of *Salmonella* infection. See Willem M. de Vos, "Fame and Future of Faecal Transplantations—Developing Next-Generation Therapies with Synthetic Microbiomes," *Microbial Biotechnology* 6, no. 4 (2013): 316–25.
11. de Vos, "Fame and Future of Faecal Transplantations," 320.
12. Colin Hill et al., "The International Scientific Association for Probiotics and Prebiotics Consensus Statement on the Scope and Appropriate Use of the Term Probiotic," *Nature Reviews Gastroenterology & Hepatology* 11, no. 8 (August 2014): 506–14, https://doi.org/10.1038/nrgastro.2014.66.
13. Patrick Veiga, Silvia Miret, and Liliana Jiménez, "Danone: The Gut Microbiome and Probiotics—100 Years of Shared History," *Nature*, Sponsor Feature, November 7, 2019, https://www.nature.com/articles/d42473-019-00336-9.
14. Jean-Marc Cavaillon and Sandra Legout, "Centenary of the Death of Elie Metchnikoff: A Visionary and an Outstanding Team Leader," *Microbes and Infection* 18, no. 10 (October 1, 2016): 577–94, https://doi.org/10.1016/j.micinf.2016.05.008.

15. Mary Ellen Sanders et al., “Probiotics and Prebiotics in Intestinal Health and Disease: From Biology to the Clinic,” *Nature Reviews Gastroenterology & Hepatology* 16, no. 10 (2019): 605–16.
16. Harry J. Flint et al., “Links between Diet, Gut Microbiota Composition and Gut Metabolism,” *Proceedings of the Nutrition Society* 74, no. 1 (2015): 13–22.
17. Sanders et al., “Probiotics and Prebiotics in Intestinal Health and Disease.”
18. Serena Sanna et al., “Causal Relationships among the Gut Microbiome, Short-Chain Fatty Acids and Metabolic Diseases,” *Nature Genetics* 51, no. 4 (April 2019): 600–605, https://doi.org/10.1038/s41588-019-0350-x.
19. Graham Rook et al., “Evolution, Human-Microbe Interactions, and Life History Plasticity,” *Lancet* 390, no. 10093 (2017): 521–30.
20. Emily R. Davenport et al., “The Human Microbiome in Evolution,” *BMC Biology* 15, no. 1 (December 27, 2017): 127, https://doi.org/10.1186/s12915-017-0454-7; Alex H. Nishida and Howard Ochman, “A Great-Ape View of the Gut Microbiome,” *Nature Reviews Genetics* 20, no. 4 (April 2019): 195–206, https://doi.org/10.1038/s41576-018-0085-z.
21. Steven A. Frese et al., “The Evolution of Host Specialization in the Vertebrate Gut Symbiont Lactobacillus Reuteri,” *PLoS Genetics* 7, no. 2 (February 17, 2011): e1001314, https://doi.org/10.1371/journal.pgen.1001314.
22. But, as pointed out above, need not be taken as evidence for a mutualistic interaction.
23. Simone Rampelli et al., “Components of a Neanderthal Gut Microbiome Recovered from Fecal Sediments from El Salt,” *Communications Biology* 4, no. 1 (February 5, 2021): 1–10, https://doi.org/10.1038/s42003-021-01689-y.
24. Nishida and Ochman, “A Great-Ape View of the Gut Microbiome.”
25. Alexandra J. Obregon-Tito et al., “Subsistence Strategies in Traditional Societies Distinguish Gut Microbiomes,” *Nature Communications* 6, no. 1 (March 25, 2015): 6505, https://doi.org/10.1038/ncomms7505.
26. D. P. Strachan, “Hay Fever, Hygiene, and Household Size,” *BMJ : British Medical Journal* 299, no. 6710 (November 18, 1989): 1259–60.
27. Blaser, *Missing Microbes.*
28. Koskella, Hall, and Metcalf, “Microbiome beyond the Horizon,” 1611.
29. Katri Korpela et al., “Intestinal Microbiome Is Related to Lifetime Antibiotic Use in Finnish Pre-School Children,” *Nature Communications* 7, no. 1 (April 2016): 10410, https://doi.org/10.1038/ncomms10410.
30. Erica D. Sonnenburg et al., “Diet-Induced Extinctions in the Gut Microbiota Compound over Generations,” *Nature* 529, no. 7585 (2016): 212–15.
31. Maria G. Dominguez-Bello et al., “Preserving Microbial Diversity,” *Sci-*

ence 362, no. 6410 (October 5, 2018): 34, https://doi.org/10.1126/science.aau8816.

32. Morin, *Community Ecology*, 348.
33. Kate E. Lynch, Emily C. Parke, and Maureen A. O'Malley, "How Causal Are Microbiomes? A Comparison with the Helicobacter Pylori Explanation of Ulcers," *Biology & Philosophy* 34, no. 6 (2019): 1–24.
34. Meghan Wymore Brand et al., "The Altered Schaedler Flora: Continued Applications of a Defined Murine Microbial Community," *ILAR Journal* 56, no. 2 (2015): 169–78.
35. Christopher E. Lawson et al., "Common Principles and Best Practices for Engineering Microbiomes," *Nature Reviews Microbiology* 17, no. 12 (2019): 725–41; Gino Vrancken et al., "Synthetic Ecology of the Human Gut Microbiota," *Nature Reviews Microbiology* 17, no. 12 (2019): 754–63.

EPILOGUE

1. See, for example, the influential paper by environmental ethicist Kenneth Goodpaster, "On Being Morally Considerable," *Journal of Philosophy* 75, no. 6 (1978): 308–25. Goodpaster writes (of plants in particular), "In the face of their obvious tendencies to maintain and heal themselves, it is very difficult to reject the idea of interests on the part of trees (and plants generally) in remaining alive" (319). The debate over this and related arguments continues. See, for example, John Basl, *The Death of the Ethic of Life* (New York: Oxford University Press, 2019).

Bibliography

Basl, John. *The Death of the Ethic of Life*. New York: Oxford University Press, 2019.

Beatty, John. "The Proximate/Ultimate Distinction in the Multiple Careers of Ernst Mayr." *Biology & Philosophy* 9, no. 3 (July 1994): 333–56. https://doi.org/10.1007/BF00857940.

Berg, Maureen, and Britt Koskella. "Nutrient-and Dose-Dependent Microbiome-Mediated Protection against a Plant Pathogen." *Current Biology* 28, no. 15 (2018): 2487–92.

Blaser, Martin. *Missing Microbes: How the Overuse of Antibiotics Is Fueling Our Modern Plagues*. Toronto: HarperCollins, 2015.

———. "Our Missing Microbes: Short-Term Antibiotic Courses Have Long-Term Consequences." *Cleveland Clinic Journal of Medicine* 85, no. 12 (December 2018): 928–30. https://doi.org/10.3949/ccjm.85gr.18005.

Boomsma, Jacobus J. "Fifty Years of Illumination about the Natural Levels of Adaptation." *Current Biology* 26, no. 24 (2016): R1250–55.

Booth, Austin. "Symbiosis, Selection, and Individuality." *Biology & Philosophy* 29, no. 5 (2014): 657–73.

Bosch, Thomas C. G., and David J. Miller. *The Holobiont Imperative*. Vienna: Springer, 2016.

Brucker, Robert M., and Seth R. Bordenstein. "The Hologenomic Basis of Speciation: Gut Bacteria Cause Hybrid Lethality in the Genus *Nasonia*." *Science* 341, no. 6146 (2013): 667–69.

Cavaillon, Jean-Marc, and Sandra Legout. "Centenary of the Death of Elie Metchnikoff: A Visionary and an Outstanding Team Leader." *Microbes and Infection* 18, no. 10 (October 1, 2016): 577–94. https://doi.org/10.1016/j.micinf.2016.05.008.

Cavicchioli, Ricardo, William J. Ripple, Kenneth N. Timmis, Farooq Azam, Lars R. Bakken, Matthew Baylis, Michael J. Behrenfeld, et al. "Scientists' Warning to Humanity: Microorganisms and Climate Change." *Nature Reviews Microbiology* 17, no. 9 (September 2019): 569–86. https://doi.org/10.1038/s41579-019-0222-5.

Clements, Frederic E. *Plant Succession; An Analysis of the Development of Vegetation*. Publication no. 242. Washington, DC: Carnegie Institution of Washington, 1916.

Costello, E. K., K. Stagaman, L. Dethlefsen, B. J. M. Bohannan, and D. A. Relman. "The Application of Ecological Theory Toward an Understanding of the Human Microbiome." *Science* 336, no. 6086 (June 8, 2012): 1255–62. https://doi.org/10.1126/science.1224203.

Darwin, Charles. *On the Origin of Species: A Facsimile of the First Edition*. Cambridge, MA: Harvard University Press, 1964.

Davenport, Emily R., Jon G. Sanders, Se Jin Song, Katherine R. Amato, Andrew G. Clark, and Rob Knight. "The Human Microbiome in Evolution." *BMC Biology* 15, no. 1 (December 27, 2017): 127. https://doi.org/10.1186/s12915-017-0454-7.

Dawkins, Richard. *The Extended Phenotype: The Long Reach of the Gene*. New York: Oxford University Press, 2016.

———. *The Selfish Gene*. New York: Oxford University Press, 1976.

Dinsmore, Charles E. *A History of Regeneration Research: Milestones in the Evolution of a Science*. Cambridge: Cambridge University Press, 2007.

Dominguez-Bello, Maria G., Kassandra M. De Jesus-Laboy, Nan Shen, Laura M. Cox, Amnon Amir, Antonio Gonzalez, Nicholas A. Bokulich, Se Jin Song, Marina Hoashi, and Juana I. Rivera-Vinas. "Partial Restoration of the Microbiota of Cesarean-Born Infants via Vaginal Microbial Transfer." *Nature Medicine* 22, no. 3 (2016): 250.

Dominguez-Bello, Maria G., Rob Knight, Jack A. Gilbert, and Martin J. Blaser. "Preserving Microbial Diversity." *Science* 362, no. 6410 (October 5, 2018): 33–34. https://doi.org/10.1126/science.aau8816.

Doolittle, W. Ford. "Is Nature Really Motherly?" *CoEvolution Quarterly* 29 (1981): 58–63.

Doolittle, W. Ford, and Austin Booth. "It's the Song, Not the Singer: An Exploration of Holobiosis and Evolutionary Theory." *Biology & Philosophy* 32, no. 1 (January 2017): 5–24. https://doi.org/10.1007/s10539-016-9542-2.

Doolittle, W. Ford, and S. Andrew Inkpen. "Processes and Patterns of Interaction as Units of Selection: An Introduction to ITSNTS Thinking." *Proceedings of the National Academy of Sciences* 115, no. 16 (April 17, 2018): 4006–14. https://doi.org/10.1073/pnas.1722232115.

Doolittle, W. Ford, and Carmen Sapienza. "Selfish Genes, the Phenotype Paradigm and Genome Evolution." *Nature* 284, no. 5757 (1980): 601–3.

Douglas, Angela E. *Fundamentals of Microbiome Science: How Microbes Shape Animal Biology*. Princeton, NJ: Princeton University Press, 2018.

Douglas, Angela E., and John H. Werren. "Holes in the Hologenome: Why Host-Microbe Symbioses Are Not Holobionts." *MBio* 7, no. 2 (2016): e02099-15. https://doi.org/10.1128/mBio.02099-15.

Doulcier, Guilhem, Amaury Lambert, Silvia De Monte, and Paul B. Rainey. "Eco-Evolutionary Dynamics of Nested Darwinian Populations and the Emergence of Community-Level Heredity." *Elife* 9 (2020): e53433. https://doi.org/10.7554/eLife.53433.

Dutreuil, Sébastien. "James Lovelock's Gaia Hypothesis: 'A New Look at Life on Earth' . . . for the Life and the Earth Sciences." In *Dreamers, Visionaries, and Revolutionaries in the Life Sciences*, edited by Oren Harman and Michael R. Dietrich, 272–87. Chicago: University of Chicago Press, 2018.

Egler, Frank Edwin. *The Nature of Vegetation, Its Management and Mismanagement: An Introduction to Vegetation Science*. Norfolk CT: Aton Forest, 1977.

Eliot, Christopher. "The Legend of Order and Chaos: Communities and Early Community Ecology." In *Philosophy of Ecology*, edited by Bryson Brown, Kevin deLaplante, and Kent A. Peacock, 11: 49–107. Handbook of the Philosophy of Science, vol. 11. Amsterdam: North-Holland, 2011.

Eme, Laura, Eleni Gentekaki, Bruce Curtis, John M. Archibald, and Andrew J. Roger. "Lateral Gene Transfer in the Adaptation of the Anaerobic Parasite Blastocystis to the Gut." *Current Biol-*

ogy 27, no. 6 (March 20, 2017): 807–20. https://doi.org/10.1016/j.cub.2017.02.003.

Engel, Philipp, Waldan K. Kwong, Quinn McFrederick, Kirk E. Anderson, Seth Michael Barribeau, James Angus Chandler, R. Scott Cornman, et al. "The Bee Microbiome: Impact on Bee Health and Model for Evolution and Ecology of Host-Microbe Interactions." *MBio* 7, no. 2 (May 4, 2016): e02164-15. https://doi.org/10.1128/mBio.02164-15.

Falkowski, Paul G., Tom Fenchel, and Edward F. Delong. "The Microbial Engines That Drive Earth's Biogeochemical Cycles." *Science* 320, no. 5879 (2008): 1034–39.

Flint, Harry J., Sylvia H. Duncan, Karen P. Scott, and Petra Louis. "Links between Diet, Gut Microbiota Composition and Gut Metabolism." *Proceedings of the Nutrition Society* 74, no. 1 (2015): 13–22.

Foster, Kevin R., Jonas Schluter, Katharine Z. Coyte, and Seth Rakoff-Nahoum. "The Evolution of the Host Microbiome as an Ecosystem on a Leash." *Nature* 548, no. 7665 (August 2017): 43–51. https://doi.org/10.1038/nature23292.

Frese, Steven A., Andrew K. Benson, Gerald W. Tannock, Diane M. Loach, Jaehyoung Kim, Min Zhang, Phaik Lyn Oh, et al. "The Evolution of Host Specialization in the Vertebrate Gut Symbiont Lactobacillus Reuteri." *PLoS Genetics* 7, no. 2 (February 17, 2011): e1001314. https://doi.org/10.1371/journal.pgen.1001314.

Gilbert, Scott F., Jan Sapp, and Alfred I. Tauber. "A Symbiotic View of Life: We Have Never Been Individuals." *Quarterly Review of Biology* 87, no. 4 (2012): 325–41.

Gleason, H. A. "The Individualistic Concept of the Plant Association." *Bulletin of the Torrey Botanical Club* 53, no. 1 (1926): 7–26. https://doi.org/10.2307/2479933.

Godfrey-Smith, Peter. "The Ant and the Steam Engine." Review of James Lovelock, *A Rough Ride to the Future*. *London Review of Books* 37, no. 4 (February 19, 2015). https://www.lrb.co.uk/the-paper/v37/n04/peter-godfrey-smith/the-ant-and-the-steam-engine?referrer=https%3A%2F%2Fwww.google.com%2F.

———. *Darwinian Populations and Natural Selection*. Oxford: Oxford University Press, 2009.

———. "Darwinism and Cultural Change." *Philosophical Transactions of the Royal Society B: Biological Sciences* 367, no. 1599 (August 5, 2012): 2160–70. https://doi.org/10.1098/rstb.2012.0118.

———. *The Philosophy of Biology*. Princeton, NJ: Princeton University Press, 2014.

Goodnight, Charles J. "Heritability at the Ecosystem Level." *Proceedings of the National Academy of Sciences* 97, no. 17 (2000): 9365–66.

Goodpaster, Kenneth E. "On Being Morally Considerable." *Journal of Philosophy* 75, no. 6 (1978): 308–25.

Goss, R. *Principles of Regeneration*. New York: Academic Press, 1969.

Hagen, Joel B. "Clementsian Ecologists: The Internal Dynamics of a Research School." *Osiris* 8 (1993): 178–95.

Hastings, P. J., Susan M. Rosenberg, and Andrew Slack. "Antibiotic-Induced Lateral Transfer of Antibiotic Resistance." *Trends in Microbiology* 12, no. 9 (September 1, 2004): 401–4. https://doi.org/10.1016/j.tim.2004.07.003.

Hehemann, Jan-Hendrik, Gaëlle Correc, Tristan Barbeyron, William Helbert, Mirjam Czjzek, and Gurvan Michel. "Transfer of Carbohydrate-Active Enzymes from Marine Bacteria to Japanese Gut Microbiota." *Nature* 464, no. 7290 (2010): 908–12.

Hehemann, Jan-Hendrik, Amelia G. Kelly, Nicholas A. Pudlo, Eric C. Martens, and Alisdair B. Boraston. "Bacteria of the Human Gut Microbiome Catabolize Red Seaweed Glycans with Carbohydrate-Active Enzyme Updates from Extrinsic Microbes." *Proceedings of the National Academy of Sciences* 109, no. 48 (2012): 19786–91.

Hendry, Andrew P. *Eco-Evolutionary Dynamics*. Princeton, NJ: Princeton University Press, 2020.

Hill, Colin, Francisco Guarner, Gregor Reid, Glenn R. Gibson, Daniel J. Merenstein, Bruno Pot, Lorenzo Morelli, et al. "The International Scientific Association for Probiotics and Prebiotics Consensus Statement on the Scope and Appropriate Use of the Term Probiotic." *Nature Reviews Gastroenterology & Hepatology* 11, no. 8 (August 2014): 506–14. https://doi.org/10.1038/nrgastro.2014.66.

Ho, Eddie K. H., and Aneil F. Agrawal. "Aging Asexual Lineages and the Evolutionary Maintenance of Sex." *Evolution* 71, no. 7 (2017): 1865–75. https://doi.org/10.1111/evo.13260.

Hooks, Katarzyna B., and Maureen A. O'Malley. "Dysbiosis and Its Discontents." *MBio* 8, no. 5 (November 8, 2017): e01492-17. https://doi.org/10.1128/mBio.01492-17.

Hull, David L. "Individuality and Selection." *Annual Review of Ecology and Systematics* 11 (1980): 311–32.

Inkpen, S. Andrew. "Health, Ecology and the Microbiome." *ELife* 8 (April 17, 2019): e47626. https://doi.org/10.7554/eLife.47626.

Inkpen, S. Andrew, Gavin M. Douglas, T. D. P. Brunet, Karl

Leuschen, W. Ford Doolittle, and Morgan G. I. Langille. "The Coupling of Taxonomy and Function in Microbiomes." *Biology & Philosophy* 32, no. 6 (December 2017): 1225–43. https://doi.org/10.1007/s10539-017-9602-2.

Jablonski, David. "Developmental Bias, Macroevolution, and the Fossil Record." *Evolution and Development* 22 (2020): 103–25. https://doi.org/10.1111/ede.12313.

———. "Scale and Hierarchy in Macroevolution." *Palaeontology* 50 (2007): 87–109. https://doi.org/10.1111/j.1475-4983.2006.00615.x.

Johnston, D. W., and E. P. Odum. 1956. "Breeding Bird Populations in Relation to Plant Succession on the Piedmont of Georgia." *Ecology* 37: 50–62.

Karam, George, Jean Chastre, Mark H. Wilcox, and Jean-Louis Vincent. "Antibiotic Strategies in the Era of Multidrug Resistance." *Critical Care* 20, no. 1 (2016): 1–9.

Kingsland, Sharon E. *The Evolution of American Ecology, 1890–2000*. Baltimore, MD: Johns Hopkins University Press, 2005.

Knapp, Dunja, Herbert Schulz, Cynthia Alexander Rascon, Michael Volkmer, Juliane Scholz, Eugen Nacu, Mu Le, Sergey Novozhilov, Akira Tazaki, and Stephanie Protze. "Comparative Transcriptional Profiling of the Axolotl Limb Identifies a Tripartite Regeneration-Specific Gene Program." *PLoS One* 8, no. 5 (2013): e61352.

Kohler, Robert E. *Landscapes and Labscapes: Exploring the Lab-Field Border in Biology*. Chicago: University of Chicago Press, 2002.

Korpela, Katri, Anne Salonen, Lauri J. Virta, Riina A. Kekkonen, Kristoffer Forslund, Peer Bork, and Willem M. de Vos. "Intestinal Microbiome Is Related to Lifetime Antibiotic Use in Finnish Pre-School Children." *Nature Communications* 7, no. 1 (April 2016): 10410. https://doi.org/10.1038/ncomms10410.

Koskella, Britt, Lindsay J. Hall, and C. Jessica E. Metcalf. "The Microbiome beyond the Horizon of Ecological and Evolutionary Theory." *Nature Ecology & Evolution* 1, no. 11 (November 2017): 1606–15. https://doi.org/10.1038/s41559-017-0340-2.

Laland, Kevin, John Odling-Smee, and John Endler. "Niche Construction, Sources of Selection and Trait Coevolution." *Interface Focus* 7, no. 5 (2017): e20160147.

Laland, Kevin N., Kim Sterelny, John Odling-Smee, William Hoppitt, and Tobias Uller. "Cause and Effect in Biology Revisited: Is Mayr's Proximate-Ultimate Dichotomy Still Useful?" *Science* 334 (2011): 6.

Land, Miriam, Loren Hauser, Se-Ran Jun, Intawat Nookaew, Michael R. Leuze, Tae-Hyuk Ahn, Tatiana Karpinets, Ole Lund, Guruprased Kora, and Trudy Wassenaar. "Insights from 20 Years of Bacterial Genome Sequencing." *Functional & Integrative Genomics* 15, no. 2 (2015): 141–61.

Langille, Morgan G. I., Jesse Zaneveld, J. Gregory Caporaso, Daniel McDonald, Dan Knights, Joshua A. Reyes, Jose C. Clemente, Deron E. Burkepile, Rebecca L. Vega Thurber, and Rob Knight. "Predictive Functional Profiling of Microbial Communities Using 16S RRNA Marker Gene Sequences." *Nature Biotechnology* 31, no. 9 (2013): 814–21.

Lawrence, Jeffrey G., and Howard Ochman. "Amelioration of Bacterial Genomes: Rates of Change and Exchange." *Journal of Molecular Evolution* 44, no. 4 (April 1, 1997): 383–97. https://doi.org/10.1007/PL00006158.

Lawrence, Jeffrey G., and John R. Roth. "Selfish Operons: Horizontal Transfer May Drive the Evolution of Gene Clusters." *Genetics* 143, no. 4 (1996): 1843–60.

Lawson, Christopher E., William R. Harcombe, Roland Hatzenpichler, Stephen R. Lindemann, Frank E. Löffler, Michelle A. O'Malley, Héctor García Martín, Brian F. Pfleger, Lutgarde Raskin, and Ophelia S. Venturelli. "Common Principles and Best Practices for Engineering Microbiomes." *Nature Reviews Microbiology* 17, no. 12 (2019): 725–41.

Lawton, John H. "Are There General Laws in Ecology?" *Oikos* 84, no. 2 (1999): 177–92.

Lean, Christopher Hunter. "Indexically Structured Ecological Communities." *Philosophy of Science* 85, no. 3 (July 2018): 501–22. https://doi.org/10.1086/697746.

Lederberg, J. "Infectious History." *Science* 288, no. 5464 (April 14, 2000): 287–93. https://doi.org/10.1126/science.288.5464.287.

Lenton, Tim, and Andrew Watson. *Revolutions That Made the Earth*. Oxford: Oxford University Press, 2013.

Levins, Richard, and Richard Lewontin. *The Dialectical Biologist*. Cambridge, MA: Harvard University Press, 1985.

Lewontin, Richard C. "The Units of Selection." *Annual Review of Ecology and Systematics*, 1970, 1–18.

Linquist, Stefan. "Against Lawton's Contingency Thesis; or, Why the Reported Demise of Community Ecology Is Greatly Exaggerated." *Philosophy of Science* 82, no. 5 (2015): 1104–15.

Losos, J. B. "Ecological Character Displacement and the Study of Adaptation." *Proceedings of the National Academy of Sciences* 97, no. 11 (May 23, 2000): 5693–95. https://doi.org/10.1073/pnas.97.11.5693.

Louca, Stilianos, Martin F. Polz, Florent Mazel, Michaeline B. N. Albright, Julie A. Huber, Mary I. O'Connor, Martin Ackermann, et al. "Function and Functional Redundancy in Microbial Systems." *Nature Ecology & Evolution* 2, no. 6 (June 2018): 936–43. https://doi.org/10.1038/s41559-018-0519-1.

Lovelock, James. *Gaia: A New Look at Life on Earth*. Oxford: Oxford University Press, 2000.

Lovelock, James E., and Lynn Margulis. "Atmospheric Homeostasis by and for the Biosphere: The Gaia Hypothesis." *Tellus* 26, no. 1–2 (1974): 2–10.

Lozupone, Catherine A., Jesse I. Stombaugh, Jeffrey I. Gordon, Janet K. Jansson, and Rob Knight. "Diversity, Stability and Resilience of the Human Gut Microbiota." *Nature* 489, no. 7415 (September 2012): 220–30. https://doi.org/10.1038/nature11550.

Lynch, Kate E., Emily C. Parke, and Maureen A. O'Malley. "How Causal Are Microbiomes? A Comparison with the Helicobacter Pylori Explanation of Ulcers." *Biology & Philosophy* 34, no. 6 (2019): 1–24.

Maienschein, Jane, and Kate MacCord. *What Is Regeneration?* Chicago: University of Chicago Press, 2022.

Margulis, Lynn. "Words as Battle Cries: Symbiogenesis and the New Field of Endocytobiology." *Bioscience* 40, no. 9 (1990): 673–77.

Martinson, Jonathan N. V., Nicholas V. Pinkham, Garrett W. Peters, Hanbyul Cho, Jeremy Heng, Mychiel Rauch, Susan C. Broadaway, and Seth T. Walk. "Rethinking Gut Microbiome Residency and the Enterobacteriaceae in Healthy Human Adults." *ISME Journal* 13, no. 9 (September 2019): 2306–18. https://doi.org/10.1038/s41396-019-0435-7.

Mayr, Ernst. "Cause and Effect in Biology." *Science* 134, no. 3489 (1961): 1501–6.

McDougall, W. B. *Plant Ecology*. Philadelphia, PA: Lea & Febiger, 1927.

McFall-Ngai, Margaret, Michael G. Hadfield, Thomas C. G. Bosch, Hannah V. Carey, Tomislav Domazet-Lošo, Angela E. Douglas, Nicole Dubilier, et al. "Animals in a Bacterial World, a New Imperative for the Life Sciences." *Proceedings of the National Academy of Sciences* 110, no. 9 (February 26, 2013): 3229–36. https://doi.org/10.1073/pnas.1218525110.

McInnes, Ross S., Gregory E. McCallum, Lisa E. Lamberte, and Willem van Schaik. "Horizontal Transfer of Antibiotic Resistance Genes in the Human Gut Microbiome." *Current Opinion in Microbiology* 53 (February 1, 2020): 35–43. https://doi.org/10.1016/j.mib.2020.02.002.

McIntosh, Robert P. "The Background and Some Current Problems of Theoretical Ecology." *Synthese* 43, no. 2 (1980): 195–255.

Medina, M., and J. L. Sachs. "Symbiont Genomics, Our New Tangled Bank." *Genomics* 95, no. 3 (2010): 129–37.

Mora, Camilo, Derek P. Tittensor, Sina Adl, Alastair G. B. Simpson, and Boris Worm. "How Many Species Are There on Earth and in the Ocean?" *PLoS Biology* 9, no. 8 (2011): e1001127.

Moran, Nancy A., and Daniel B. Sloan. "The Hologenome Concept: Helpful or Hollow?" *PLoS Biology* 13, no. 12 (2015): e1002311.

Morin, Peter J. *Community Ecology*. 2nd ed. Hoboken, NJ: John Wiley & Sons, 2011.

Mueller, Noel T., Elizabeth Bakacs, Joan Combellick, Zoya

Grigoryan, and Maria G. Dominguez-Bello. "The Infant Microbiome Development: Mom Matters." *Trends in Molecular Medicine* 21, no. 2 (2015): 109–17.

Nalepa, C. A., D. E. Bignell, and C. Bandi. "Detritivory, Coprophagy, and the Evolution of Digestive Mutualisms in Dictyoptera." *Insectes Sociaux* 48, no. 3 (2001): 194–201.

Nicholson, Malcolm. "Henry Allan Gleason and the Individualistic Hypothesis: The Structure of a Botanist's Career," *Botanical Review* 56 (1990): 91–161.

Nishida, Alex H., and Howard Ochman. "A Great-Ape View of the Gut Microbiome." *Nature Reviews Genetics* 20, no. 4 (April 2019): 195–206. https://doi.org/10.1038/s41576-018-0085-z.

Nyholm, Spencer V., and Margaret McFall-Ngai. "The Winnowing: Establishing the Squid-Vibrio Symbiosis." *Nature Reviews Microbiology* 2, no. 8 (2004): 632–42.

Obregon-Tito, Alexandra J., Raul Y. Tito, Jessica Metcalf, Krithivasan Sankaranarayanan, Jose C. Clemente, Luke K. Ursell, Zhenjiang Zech Xu, et al. "Subsistence Strategies in Traditional Societies Distinguish Gut Microbiomes." *Nature Communications* 6, no. 1 (March 25, 2015): 6505. https://doi.org/10.1038/ncomms7505.

Odenbaugh, Jay. "Seeing the Forest and the Trees: Realism about Communities and Ecosystems." *Philosophy of Science* 74, no. 5 (2007): 628–41.

Ogden, Daniel. *Drakon: Dragon Myth and Serpent Cult in the Greek and Roman Worlds*. Oxford: Oxford University Press, 2013.

Okasha, Samir. *Evolution and the Levels of Selection*. Oxford: Oxford University Press, 2006.

Oliphint, Paul A., Naila Alieva, Andrea E. Foldes, Eric D. Tytell, Billy Y.-B. Lau, Jenna S. Pariseau, Avis H. Cohen, and Jennifer R. Morgan. "Regenerated Synapses in Lamprey Spinal Cord Are Sparse and Small Even after Functional Recovery from Injury." *Journal of Comparative Neurology* 518, no. 14 (2010): 2854–72.

Oliver, Kerry M., Patrick H. Degnan, Gaelen R. Burke, and Nancy A. Moran. "Facultative Symbionts in Aphids and the Horizontal Transfer of Ecologically Important Traits." *Annual Review of Entomology* 55, no. 1 (2010): 247–66. https://doi.org/10.1146/annurev-ento-112408-085305.

O'Malley, Maureen. "'Everything Is Everywhere: But the Environment Selects': Ubiquitous Distribution and Ecological Determinism in Microbial Biogeography." *Studies in History and Philosophy of Science Part C: Studies in History and Philosophy of Biological and Biomedical Sciences* 39, no. 3 (September 2008): 314–25. https://doi.org/10.1016/j.shpsc.2008.06.005.

———. "The Nineteenth Century Roots of 'Everything Is Everywhere.'" *Nature Reviews Microbiology* 5, no. 8 (August 2007): 647–51. https://doi.org/10.1038/nrmicro1711.

———. *Philosophy of Microbiology*. Cambridge: Cambridge University Press, 2014.

Orgel, Leslie E., and Francis H. C. Crick. "Selfish DNA: The Ultimate Parasite." *Nature* 284, no. 5757 (1980): 604–7.

Pennekamp, Frank, Mikael Pontarp, Andrea Tabi, Florian Altermatt, Roman Alther, Yves Choffat, Emanuel A. Fronhofer, Pravin Ganesanandamoorthy, Aurélie Garnier, and Jason I. Griffiths. "Biodiversity Increases and Decreases Ecosystem Stability." *Nature* 563, no. 7729 (2018): 109–12.

Perry, David A. *Forest Ecosystems*. Baltimore, MD: Johns Hopkins University Press, 1994.

Philippot, Laurent, Bryan S. Griffiths, and Silke Langenheder. "Microbial Community Resilience across Ecosystems and Multiple Disturbances." *Microbiology and Molecular Biology Reviews* 85, no. 2 (2021): e00026-20.

Quammen, David. *The Tangled Tree: A Radical New History of Life*. New York: Simon and Schuster, 2018.

Rampelli, Simone, Silvia Turroni, Carolina Mallol, Cristo Hernandez, Bertila Galván, Ainara Sistiaga, Elena Biagi, et al. "Components of a Neanderthal Gut Microbiome Recovered from Fecal Sediments from El Salt." *Communications Biology* 4, no. 1 (February 5, 2021): 1–10. https://doi.org/10.1038/s42003-021-01689-y.

Rook, Graham, Fredrik Bäckhed, Bruce R. Levin, Margaret J. McFall-Ngai, and Angela R. McLean. "Evolution, Human-Microbe Interactions, and Life History Plasticity." *Lancet* 390, no. 10093 (2017): 521–30.

Rosenberg, Eugene. *Microbiomes: Current Knowledge and Unanswered Questions*. Cham: Springer International Publishing, 2021. https://doi.org/10.1007/978-3-030-65317-0.

Rosenberg, Eugene, and Ilana Zilber-Rosenberg. *The Hologenome Concept: Human, Animal and Plant Microbiota*. Berlin: Springer, 2014.

———. "The Hologenome Concept of Evolution after 10 Years." *Microbiome* 6, no. 1 (2018): 1–14.

Roughgarden, Joan, Scott F. Gilbert, Eugene Rosenberg, Ilana Zilber-Rosenberg, and Elisabeth A. Lloyd. "Holobionts as Units of Selection and a Model of Their Population Dynamics and Evolution." *Biological Theory* 13, no. 1 (2018): 44–65.

Ruse, Michael. *The Gaia Hypothesis: Science on a Pagan Planet.* Chicago: University of Chicago Press, 2013.

Sanders, Mary Ellen, Daniel J. Merenstein, Gregor Reid, Glenn R. Gibson, and Robert A. Rastall. "Probiotics and Prebiotics in Intestinal Health and Disease: From Biology to the Clinic." *Nature Reviews Gastroenterology & Hepatology* 16, no. 10 (2019): 605–16.

Sanna, Serena, Natalie R. van Zuydam, Anubha Mahajan, Alexander Kurilshikov, Arnau Vich Vila, Urmo Võsa, Zlatan Mujagic, et al. "Causal Relationships among the Gut Microbiome, Short-Chain Fatty Acids and Metabolic Diseases." *Nature Genetics* 51, no. 4 (April 2019): 600–605. https://doi.org/10.1038/s41588-019-0350-x.

Schneider, Stephen Henry, and Randi Londer. *Coevolution of Climate and Life.* San Francisco: Sierra Club Books, 1984.

Sonnenburg, Erica D., Samuel A. Smits, Mikhail Tikhonov, Steven K. Higginbottom, Ned S. Wingreen, and Justin L. Sonnenburg. "Diet-Induced Extinctions in the Gut Microbiota Compound over Generations." *Nature* 529, no. 7585 (2016): 212–15.

Sonnenburg, Justin L. "Microbiome Engineering." *Nature* 518, no. 7540 (February 26, 2015): S10. https://doi.org/10.1038/518S10a.

Sonnenburg, Justin, and Erica Sonnenburg. *The Good Gut: Taking Control of Your Weight, Your Mood, and Your Long-Term Health.* New York: Penguin Books, 2016.

Steffen, Will, Katherine Richardson, Johan Rockström, Hans Joachim Schellnhuber, Opha Pauline Dube, Sébastien Dutreuil, Timothy M. Lenton, and Jane Lubchenco. "The Emergence and Evolution of Earth System Science." *Nature Reviews Earth & Environment* 1, no. 1 (2020): 54–63.

Sterelny, Kim. "Local Ecological Communities." *Philosophy of Science* 73, no. 2 (April 2006): 215–31. https://doi.org/10.1086/510819.

Strachan, D. P. "Hay Fever, Hygiene, and Household Size." *BMJ: British Medical Journal* 299, no. 6710 (November 18, 1989): 1259–60.

Stuart, Yoel E., S. Andrew Inkpen, Robin Hopkins, and Daniel I. Bolnick. "Character Displacement Is a Pattern: So, What Causes It?" *Biological Journal of the Linnean Society* 121, no. 3 (July 2017): 711–15. https://doi.org/10.1093/biolinnean/blx013.

Theis, Kevin R., Nolwenn M. Dheilly, Jonathan L. Klassen, Robert M. Brucker, John F. Baines, Thomas C. G. Bosch, John F. Cryan, Scott F. Gilbert, Charles J. Goodnight, and Elisabeth A. Lloyd. "Getting the Hologenome Concept Right: An Eco-Evolutionary Framework for Hosts and Their Microbiomes." *Msystems* 1, no. 2 (2016): e00028-16.

Turnbaugh, Peter J., Micah Hamady, Tanya Yatsunenko, Brandi L. Cantarel, Alexis Duncan, Ruth E. Ley, Mitchell L. Sogin, et al. "A Core Gut Microbiome in Obese and Lean Twins." *Nature* 457, no. 7228 (January 2009): 480–84. https://doi.org/10.1038/nature07540.

Tyler, Toby, *On Gaia: A Critical Investigation of the Relationship between Life and Earth*. Princeton, NJ: University of Princeton Press, 2013.

Van Vliet, Simon, and Michael Doebeli. "The Role of Multilevel Selection in Host Microbiome Evolution." *Proceedings of the National Academy of Sciences* 116, no. 41 (2019): 20591–97.

Veiga, Patrick, Silvia Miret, and Liliana Jiménez. "Danone: The Gut Microbiome and Probiotics—100 Years of Shared History." *Nature*, Sponsor Feature, November 7, 2019. https://www.nature.com/articles/d42473-019-00336-9.

Vellend, Mark. "Conceptual Synthesis in Community Ecology." *Quarterly Review of Biology* 85, no. 2 (June 2010): 183–206. https://doi.org/10.1086/652373.

———. *The Theory of Ecological Communities*. Princeton, NJ: Princeton University Press, 2016.

Vergara, M. Natalia, George Tsissios, and Katia Del Rio-Tsonis. "Lens Regeneration: A Historical Perspective." *International Journal of Developmental Biology* 62, no. 6–7–8 (2018): 351–61. https://doi.org/10.1387/ijdb.180084nv.

Vernier, Cassondra L., Iris M. Chin, Boahemaa Adu-Oppong, Joshua J. Krupp, Joel Levine, Gautam Dantas, and Yehuda Ben-Shahar. "The Gut Microbiome Defines Social Group Membership in Honey Bee Colonies." *Science Advances* 6, no. 42 (October 1, 2020): eabd3431. https://doi.org/10.1126/sciadv.abd3431.

Vos, Willem M. de. "Fame and Future of Faecal Transplantations—Developing Next-Generation Therapies with Synthetic Microbiomes." *Microbial Biotechnology* 6, no. 4 (2013): 316–25.

Vrancken, Gino, Ann C. Gregory, Geert R. B. Huys, Karoline Faust, and Jeroen Raes. "Synthetic Ecology of the Human Gut Microbiota." *Nature Reviews Microbiology* 17, no. 12 (2019): 754–63.

Whitham, Thomas G., Gerard J. Allan, Hillary F. Cooper, and Stephen M. Shuster. "Intraspecific Genetic Variation and Species Interactions Contribute to Community Evolution." *Annual Review of Ecology, Evolution, and Systematics* 51, no. 1 (2020): 587–612. https://doi.org/10.1146/annurev-ecolsys-011720-123655.

Whitham, Thomas G., Joseph K. Bailey, Jennifer A. Schweitzer, Stephen M. Shuster, Randy K. Bangert, Carri J. LeRoy, Eric V. Lonsdorf, et al. "A Framework for Community and Ecosystem Genetics: From Genes to Ecosystems." *Nature Reviews Genetics* 7, no. 7 (July 2006): 510–23. https://doi.org/10.1038/nrg1877.

Wollenberg, Michael S., and Edward G. Ruby. "Phylogeny and Fitness of *Vibrio fischeri* from the Light Organs of *Euprymna scolopes* in Two Oahu, Hawaii, Populations." *ISME Journal* 6, no. 2 (2012): 352–62.

Worster, Donald. *Nature's Economy: A History of Ecological Ideas.* 2nd ed. New York: Cambridge University Press, 1994.

Wymore Brand, Meghan, Michael J. Wannemuehler, Gregory J. Phillips, Alexandra Proctor, Anne-Marie Overstreet, Albert E. Jergens, Roger P. Orcutt, and James G. Fox. "The Altered Schaedler Flora: Continued Applications of a Defined Murine Microbial Community." *ILAR Journal* 56, no. 2 (2015): 169–78.

Yong, Ed. *I Contain Multitudes: The Microbes within Us and a Grander View of Life*. New York: Ecco, HarperCollins, 2016.

Zilber-Rosenberg, Ilana, and Eugene Rosenberg. "Role of Microorganisms in the Evolution of Animals and Plants: The Hologenome Theory of Evolution." *FEMS Microbiology Reviews* 32, no. 5 (2008): 723–35.

Index

Page numbers in italics refer to figures.